THE GOLDEN HELIX

INSIDE BIOTECH VENTURES

THE GOLDEN HELIX

INSIDE BIOTECH VENTURES

Arthur Kornberg

UNIVERSITY SCIENCE BOOKS
Sausalito, California

University Science Books
55D Gate Five Road
Sausalito, CA 94965
www.uscibooks.com

Production manager: *Susanna Tadlock*
Manuscript editor: *Gunder Hefta*
Designer: *Robert Ishi*
Illustrator: *Georg Klatt*
Compositor: *Wilsted & Taylor*
Printer and Binder: *Maple-Vail Book Manufacturing Group*

This book is printed on acid-free paper.

Library of Congress as Cataloged

Kornberg, Arthur, 1918–
      The golden helix : inside biotech ventures / Arthur Kornberg.
         p.   cm.
      Includes bibliographical references and index.
      ISBN 0-935702-32-6 (cloth). — ISBN 1-891389-19-X (pbk.)
      1. Biotechnology industries — United States — Case studies. 2. New
business enterprises — United States — Case studies. 3. Biotechnology —
United States — Industrial applications — Case studies.
4. Biotechnologists — United States.   I. Title.
HD9999.B443U647 1995
338.7'6606'0973—dc20 1995   94–43482
                                                      CIP

Necessity is seldom the mother of invention.
Rather, true inventions beget necessities.

Contents

Preface   ix

CHAPTER 1
Currents and Eddies in Biotechnology   1

Biotech Ventures   2
Currents and Eddies in Science   3
Fashions in Research   5
Biology or Technology: Academia versus Industry   9
The Two Cultures: Chemistry versus Biology
in Academia and in the Pharmaceutical Industry   13

CHAPTER 2
Birth of a Biotech Venture   19

My Path to Biotech Ventures   19
Conception of DNAX   27
Infancy   29
The Laboratory   31
Scientific Staff   31
Scientific Advisory Board   47
The Science Plan   50
The Business Plan   54

CHAPTER 3
A Scientist–Entrepreneur: Alejandro Zaffaroni   59

Youth in Uruguay   59
Graduate Study in Rochester   62
Scientist–Entrepreneur at Syntex   65
Turning Syntex into a Pharmaceutical Company   69
Founding of ALZA   73
From the Brink of Bankruptcy   84
Defeat at Dynapol   91
Lure of Biotechnology: DNAX and Affymax   93
Attitudes and Philosophies   94

CHAPTER 4

A Biotech-Driven Pharmaceutical Company   99

Origins of Schering-Plough Corporation   100
Robert P. Luciano   107
Hugh D'Andrade   118
Acquisition of DNAX   121

CHAPTER 5

Growth of a Biotech Venture   131

The Early Golden Years: Dawn of the Cytokines   132
A Trough between Waves   141
The DNAX Route to Academia   151

CHAPTER 6

The DNAX–Schering-Plough Connection   159

The DNAX–Schering-Plough Connection: Past, Present, and Future   159
Schering-Plough Middle Management   164
The DNAX Founders   172
The DNAX Future   186

CHAPTER 7

Genentech, Amgen, Chiron, and Regeneron   195

Genentech, Biotech Pioneer   196
Amgen, with Blockbuster Drugs   202
Chiron, with Protean Goals   209
Regeneron, Pioneer in Neuroscience   216

CHAPTER 8

Pros and Cons of Biotech Ventures   231

Secrecy and Patents   231
The Cetus Patent for PCR   236
The Stanford–UC Patent for Recombinant DNA   241
Biotech Ventures: Pros and Cons   245
Lessons from Biotech Ventures   248
The Future of Biotech Ventures   254

Epilogue   259
References   261
Glossary   263
Index   275

Preface

Genetic engineering and associated technologies have brought about the most revolutionary advances in the history of biological and medical science. Applications of this genetic and chemical knowledge created a biotechnology industry with vast economic and social potential. The biologists and biochemists who invented these new technologies in their academic laboratories came to the forefront of entrepreneurial ventures to use this knowledge to develop drugs and devices for the diagnosis, prevention, and treatment of disease.

Along with other biologists and biochemists engaged in basic research who previously had shunned all commercial connections, I found myself attracted by the resources that industry could provide to apply the novel techniques we all helped to discover. The twenty-five years I spent finding the enzymes that make DNA in our cells supplied the reagents for others to create recombinant DNAs and to trigger an avalanche of ingenious ways to use these techniques to advance basic knowledge in biology, as well as to devise highly useful industrial products and procedures.

Of the thousand or more biotech ventures in the United States, virtually all were financed by venture capitalists and investors whose interest in research was to create a profitable business rather than to acquire knowledge for its own sake. There was little sympathy for the usually long and always unpredictable time scale required for innovative discoveries. With Alejandro Zaffaroni, a scientist and intimate friend, who for three decades had successfully applied his entrepreneurial skills in the pharmaceutical industry, I could share the faith that creative scientists at a frontier of medical science would, in due course, make novel discoveries worthy of industrial development. We were joined by Paul Berg and Charles Yanofsky—Stanford colleagues, friends, and innovators in biotechnology—who had the same dedication to basic science. Together we

ix

founded the DNAX Institute of Molecular and Cellular Biology, Inc., in 1980.

The Schering-Plough Corporation, a medium-size pharmaceutical company based in New Jersey, acquired DNAX a little more than a year later, a move directed by Robert Luciano, its CEO. The acquisition of DNAX was based on his conviction that the industry would become biotechnology-driven and that a venture both strong on science and located outside the company's establishment in New Jersey was needed to provide the means for Schering-Plough to achieve that transformation. DNAX, in turn, found in Schering-Plough an ideal patron with an understanding of the style and time scale required for basic research.

In this narrative, which features the DNAX–Schering-Plough partnership, I describe how it made DNAX a world leader in basic immunology and at the same time generated for Schering-Plough multiple candidates for drug development far earlier than expected. Because the success of this venture depended so much on the people on both sides—academic and industrial—emphasis will be placed on their personalities and how they bridged the cultures and missions of their different worlds.

Beyond DNAX and other biotech ventures (Genentech, Amgen, Chiron, and Regeneron) to be described more briefly, I reflect on some general issues that affect the conduct of science—in particular, liaisons and conflicts between academia and industry. I consider the pros and cons of biotech ventures, secrecy, patents, and the current oxymoron: targeted (or strategic) basic research. From my varied experiences, I find a renewed confidence in the power of science to allow us to understand ourselves and our world in rational, molecular terms and in the capacity of motivated people to apply this knowledge in practical ways to improve the quality of our lives.

In 1994, the outlook for biotech ventures turned bleak. Obituaries were being written for an industry overexpanded by greed for profit and unmindful of the slow and tortuous course from discovery to a marketable product. Turning the DNA helix into gold seemed another alchemist dream. We can expect that, after an inevitable shakedown, mature biotech ventures will exploit the new science and will evolve technologies to shape the chemistry that defines our state of body and mind well into the next millennium.

To those who gave their time for interviews and to those who read drafts of various sections of the book, I am most grateful. I am especially indebted to the many who expended great effort in reading the entire manuscript and then made highly useful suggestions. These include Jane Ellis, Hugh D'Andrade, Spyros Andreopoulos, Carol Dempster, Horace Judson, Charlene Kornberg, Ken Kornberg, Roger Kornberg, Tom Kornberg, Robert Lehman, Paul Lubetkin, Paul Schimmel, W. Denman Van Ness, and Charles Yanofsky. I offer special thanks to Paula Hagen at Schering-Plough and Constance Mitchell at ALZA for fact-checking and helpful suggestions. To Beverly Forsyth, who typed the manuscript from handwritten copy; to Bruce Armbruster, friend and publisher; to Gunder Hefta, editor; and to Robert Ishi, book designer, I give my heartfelt thanks.

Arthur Kornberg

THE GOLDEN HELIX

INSIDE BIOTECH VENTURES

CHAPTER 1

Currents and Eddies
in Biotechnology

At the end of 1993, there were 1272 biotechnology companies in the United States. These companies employed 80,000 people directly, and they employed at least again as many in the construction of laboratories, the publication of news reports, the manufacturing and distribution of reagents, and in a variety of other activities, including Wall Street banking and analysis. Biotechnology, with annual sales in 1993 of $6 billion (a 35 percent increase over the previous year) is expected to reach five to ten times that volume by the year 2000. The phenomenal growth of the industry is fueled by investors, academic institutions, politicians, and the hopes of many millions who suffer from incurable diseases, for which, they trust, the industry will soon find a cure.

The truly revolutionary developments in medicine and biology in recent decades are based on genetic engineering and its associated biotechnologies and (although this is less widely appreciated) on the coalescence of the medical and biologic sciences into a unified discipline whose messages are expressed through the language of chemistry.

Revolutionary, an overused word, is justifiably applied to these extraordinary movements in science. Another development that is surely astonishing, if not revolutionary, is the extensive involvement of biologists and biochemists in entrepreneurial activities, something that was utterly unimaginable only a decade or so back.

1

Biotech Ventures

Because the techniques of the biotechnology explosion were created entirely by academic scientists, virtually all biotech ventures in the United States include biologists and biochemists as founders, managers, or advisors. In most vigorous university departments of the medical and biologic sciences, prominent faculty members have one or several industrial connections. I never thought that, in 1968, after more than 25 years of full-time academic research in biochemistry, having avoided all commercial associations, I would become an advisor to a new biotech venture. Nor did I believe that, in the next 25 years, I would be a founder, advisor, or director of a half dozen more. Such affiliations entail advantages and threats to the progress of science and the betterment of human welfare that must be analyzed in order to be understood, and be understood in order to guide our actions.

With the prospects of novel drugs and procedures for medicine and agriculture, major pharmaceutical companies were spurred to join in some of these ventures, as well as to initiate similar programs of their own. In many cases, the relations between the corporate giants and the fledgling enterprises were awkward and stormy. In at least one instance, however, the match worked to create a symbiosis of world-class basic research that generated a pipeline of products for drug development.

Viewed from ten years after the fact, the acquisition of the DNAX Research Institute by the Schering-Plough Corporation is a story of how personalities, philosophies, and sustained policies combined to make this union a unique success. In this chronicle, I will reflect upon these people, their backgrounds, their attitudes, and their actions, all of which were essential to blending the alien cultures of academia and big business. These people—scientists, advisors, and managers, both academic and industrial—gave DNAX the constant devotion that was needed to keep it nourished and on course.

On the DNAX side, the entrepreneurial drive came from Alex Zaffaroni. Trained in biochemistry and endocrinology, and with extensive pharmaceutical experience in the applications of scientific discoveries to medical uses, he saw an industrial potential for developments in molecular biology in the 1960s and 1970s. Paul Berg, Charles Yanofsky, and I, who had contributed significantly to these developments, had been turned off by overtures from venture capitalists, but we felt that, with Zaffaroni, we could create an enterprise in which pursuit of basic research with a communal focus could be linked to the development of drugs for treating disease.

On the Schering-Plough side, Robert Luciano, the chief executive officer, was captivated by the notion of biotechnology as the frontier of the pharmaceutical industry. He regarded the acquisition of DNAX as a means to drive Schering-Plough to that frontier. He set no timetable, and he had no illusions about short-term returns on the investment. To Hugh D'Andrade, the director of strategic planning, he entrusted the arrangements for this novel academic enterprise, which was culturally and geographically remote from the company's New Jersey operations.

On both sides, friendships were cultivated and lessons were learned. DNAX founders and scientists came to respect the skills needed for drug development, manufacture, regulatory approval, and marketing. Schering-Plough managers and scientists became aware of the haphazard nature of discovery, the erratic pace of research progress, and the futility of establishing milestones. Mutual respect and affection at the top level resolved misunderstandings, maintained adherence to academic standards, and kept the long-range goals in sight.

The focus in this narrative will be on DNAX and its Schering-Plough sponsor not only because I know them best but also because, in this joint biotech venture, I found principles and practices that merit attention and emulation. Inasmuch as other successful ventures in biotechnology have differed widely from DNAX in their conception, development, and personality, I have included brief accounts of some of them—Genentech, Amgen, Chiron, and Regeneron. In every case, biotech ventures must confront profound problems in the general conduct of science, the waves of fashion that move research, conflicts between academia and industry, and the need to bridge the age-old gap between the cultures of chemistry and biology.

Currents and Eddies in Science

River metaphors offer a framework for describing the progress of history and science: Brooks grow to streams that merge to form mighty rivers. The once discrete medical disciplines of anatomy, physiology, biochemistry, bacteriology, pharmacology, and pathology have coalesced into a single, powerful flow of knowledge that can be expressed in the universal language of chemistry. So too have the emerging biological disciplines of genetics, cell biology, developmental biology, and molecular biology. At times, forces obstruct the flow, creating eddies and diverting it from its main course, which is approaching ever closer to the secrets of nature. So it has been with the flow of the medical and biological sciences in this century.

Ryszard Kapuscinski, a Polish journalist, said of the French historian Fernand Braudel: "He wrote that history is like a river. On the surface it flows rapidly and disappears. But down below, there is a deep stream which moves more slowly, doesn't change quickly, but is the more important because it drives the whole river." What interests us all is finding that deep current—in my case, the most rational understanding of life: its reduction to the molecular details of chemistry.

Anatomy, once the most descriptive of the life sciences, is now understandable in terms of the assembly of macromolecules to form the organelles, cells, and tissues of the organism. Genetics, only a few decades ago the most abstract, has been reduced to simple genetic chemistry. These two disciplines, previously at the extremes, have intersected. Embryology and genetics have become indistinguishable in their mission to identify the temporal and spatial expression of genes that encode the traits of each individual of a species. All these disciplines offer a variety of approaches with the same end in view: an understanding of the molecular basis of growth and senescence and of health and disease—and, therefore, an understanding of how best to intervene in order to forestall and correct the aberrations caused by genetic deficiencies and environmental stresses.

Like the geographical features that arrest or alter the currents of rivers, human and societal factors may divert or completely dam the progress of biological and medical science. A gulf separates the cultures of chemistry and biology; fashions in research lead to the abandonment of fertile subjects; scientific curiosity is discouraged when it may seem irrelevant to urgent, practical needs; confusions between biology and technology—confusions in the minds of academics, industrialists, and the public—pit academia against industry; pressures are exerted to meet milestones set for industrial targets irrespective of the deeper pursuit of knowledge for its own sake; secrecy and espionage stifle scientific pursuits; unfounded and debilitating patent litigation beclouds an era of the most profound scientific and social change and progress. All this turbulence leads people to question the concept of progress, and it invites politicians, ideologues, and journalists to arouse fear and to spread misinformation about science and biotechnology.

Fashions in Research

Fashions prevail in research as in all other departments of human behavior; tides erode one beach as they create another. The hunting metaphor introduced by Paul de Kruif in his *Microbe Hunters* aptly describes the succession of movements in medical science in this century. Those hunters, who in the first two decades found the microbes responsible for many of the scourges of mankind, were replaced in the next two decades by the *vitamin hunters*, who discovered that deficiencies of vitamins could cause other epidemic diseases—pellagra, rickets, and scurvy. The vitamin hunters were superseded by the *enzyme hunters*, who showed how enzymes were assisted by vitamins in the metabolic operations responsible for cell growth and energy metabolism.

In recent decades, the *gene hunters* have dominated. With an inexhaustible supply of genes and the capacity to manipulate them in the minutest detail, the gene hunters have shown that species can be modified at their will. Bacterial, plant, and animal factories are created to produce massive quantities of the rarest hormones, cytokines, and antibodies for medicine, agriculture, and industry. A cadre of gene hunters is busy looking for the genes for human traits and the defects in these genes that cause inborn errors of metabolism. But the searches by these disease hunters are not basically different from those established for bacteria, fungi, plants, flies, and other animals. The greatest mystery now resides in brain processes—mood, memory, mental illness—which, when probed successfully with novel technologies, will turn the spotlight on a new breed of hunters (to whom we might refer as *head hunters*).

As observed by the Swiss physicist–philosopher Marcus Fierz, the scientific insights of an age can shed such glaring light on an area as to leave the rest in even greater darkness. The incandescence of enzymology was so dazzling that attention to nutrition as a science faded nearly to the vanishing point, leaving the major questions of human nutrition unattended. Why do deficiencies in vitamins, which are needed by every cell in the body, cause diarrhea, dermatitis, and dementia in the case of niacin and neuritis in the case of thiamine? The refined knowledge of biochemical mechanisms has obscured our gaping ignorance about the physiology of cells and the organism. For lack of such basic information, the science of nutrition is in a sorry state and controversies rage over what and how much to eat. With widespread concerns about the value or avoidance of cholesterol, sugar, salt, fiber, fat, and megadose vitamins, the only clear winners are the zealots and quacks.

Just as enzymology eclipsed nutrition, so has genetic engineering, with its mastery over DNA, cast a shadow over enzymology. To the current generation of molecular biologists, enzymes come in kits and are as faceless as buffers and salts. Yet, for lack of attention to enzymes, the truly molecular basis of biology will remain obscure; profound questions of how cells and organisms function and develop will not be answered. How is the chromosome rearranged by enzymes to produce the genes for antibodies? What enzymes direct embryonic tissues to become specialized adult organs? How do the enzymatic processes of growth and senescence proceed in response to different programs and circumstances?

Fashions prevail within, as well as among, disciplines. The enormous research effort to prevent and treat AIDS (acquired immune deficiency syndrome) has increased interest in the structure and life cycle of HIV (human immunodeficiency virus, the causative agent of AIDS). Yet with all this clearly important activity, the ways in which this and other viruses enter cells and are uncoated to expose their genomes for replication are largely ignored. Studies of bacterial viruses, as well as those of animal viruses, long ago identified a particular surface protein on each virus, the "adsorption protein," and the structure on the host-cell surface to which each attaches, the "receptor." Once they were named, little more was done to learn their molecular details and how they operate. Further, it remains a mystery how the viral chromosome (RNA or DNA), upon entering the cell, instantly appropriates key elements of the cellular replication machinery in competition with a host genome a thousand or more times its size. Knowing more about these operations is bound to be of wide significance and of invaluable help in coping with the many viral infections that will continue to plague us.

As scientific explorations overlap one another, they widen our knowledge base, but they also inevitably increase the perimeter of our ignorance. In the pursuit of the unknown, the deployment of resources is influenced more by social, economic, and political forces than by curiosity and inspiration. Fertile fields are left fallow as scientists cluster in fashionable areas for the security of being a part of a popular movement.

Few scientists have the courage, confidence, or independence to pursue a problem that appears irrelevant to their colleagues or that lacks a practical objective. It may seem, even to many scientists, unreasonable and impractical (call it *counterintuitive*) to address an urgent problem, such as a life-threatening disease, by pursuing apparently unrelated questions in basic biology or chemistry. Yet, the pursuit of curiosity about the basic facts of nature has proven, throughout the history of medical science, to be the most practical and the most *cost-effective* route to successful drugs and devices.

In an often retold parable, a surgeon, while jogging around a lake, spotted a man drowning. He dove in, dragged the victim ashore, and resuscitated him. His duty done, he wearily resumed his jogging, only to see two more people drowning. He also saw a colleague, a professor of biochemistry, standing nearby, apparently absorbed in thought. He called to him to go after one while he went after the other. When the biochemist was slow to respond, the surgeon shouted, "Why don't you do something?" The biochemist said, "I *am* doing something. I'm trying to figure out who's throwing all these people in the lake."

This parable is not intended to convey a lack of regard for fundamental issues among physicians nor a callousness among scientists. Rather, it is meant to portray the reality that a serious problem, such as a war on disease, is often best fought on several fronts. Some contribute their special skills to the distressed individual, while others try to gain the breadth of knowledge necessary to outwit both present and future enemies.

Investigations that seemed totally irrelevant to any practical objective have yielded most of the major discoveries of medicine: x rays, from a physicist observing electrical discharges in a vacuum tube; penicillin, from enzyme studies of bacterial lysis; the polio vaccine, from learning how to grow cells in culture; nylon and neoprene, from showing that polymers are linked by the same forces as small molecules; and the discovery of genetic engineering and recombinant DNA, from exploring DNA biochemistry. All these discoveries have come from the pursuit of curiosity about questions in physics, chemistry, and biology, apparently unrelated at the outset to a specific medical or practical problem. In this vein, I will venture a prediction that the cure for AIDS will come from basic studies of the immune system rather than from screening for drugs and vaccines.

A meeting in Washington, D.C., in 1990 celebrated the bicentennial of the United States Patent and Trademark Office. From discussions among the inventors and the corporate and government officials gathered there, a remarkable truth emerged. It was generally agreed that the age-old saying "necessity is the mother of invention" is usually wrong. Generally, the reverse has proven to be true: *invention is the mother of necessity*. Inventions only later become necessities.

Time and again, inventors have created things that had to wait for many years to be recognized for their practical value. Nobody really needed the airplane, the FM radio, television, xerography, the lasers that gave us the bar code and compact disc, or the quantum mechanics that led to the transistor. It took Chester Carlson, the inventor of xerography or electrostatic printing (more familiarly known as the Xerox process) six years to interest a company in his invention, and it was twenty years before the first commercial xerographic copier was produced. The Englishman and the German who independently invented the jet-propelled airplane in the early 1930s were unable to interest companies, or the air forces in either of their countries, in developing their invention. Fax machines were invented 30 years ago, but it took a deteriorating postal service (among other factors) to make them the necessities they are today.

Even industrial inventions emerge from creative processes, which are haphazard rather than goal-oriented. The process of invention conflicts with prudent business strategy. A pioneering invention is, almost by definition, profoundly different from what a company has been doing routinely. Because such an invention is commercially unproved, it is therefore seen to be riskier than the established business.

The lessons to be learned from this history should be crystal clear: It is crucial for a society, a culture, or a company to understand the nature of the creative process and to provide for its support. No matter how counterintuitive this may seem, basic research is the lifeblood of practical advances in medicine, just as pioneering inventions are the source of industrial strength. The future is invented, not predicted. To paraphrase Clarence Dutton (a geologist, explorer, and chronicler of the American West), great innovations, whether in art, literature, or science, seldom take the world by storm. They must be cultivated to be understood, and they must be understood before they can be estimated.

Of course, it is important that the basic discoveries be promptly and wisely applied to solve practical problems. The recent applications of biotechnology to medicine have given us major insights into diabetes, cancer, and other metabolic diseases, and these techniques should be equally effective when applied to investigations of the brain and behavior.

Biology or Technology: Academia versus Industry

Biotechnology seems a less frightening term than *genetic engineering*, which may conjure up thoughts of misguided or malevolent attempts to clone microbial or human monsters. *Engineering*, when applied to genetics, has, to many people, a bad ring that is absent for engineering designated as mechanical, electrical, civil, or chemical. Recombinant DNA and broadly associated technologies are seen by the public as more acceptable when they are referred to as biotechnology. Yet the adoption of this term carries a danger: the blurring of an important distinction between biology, which is a quest for knowledge about life, and technology, which is a means to apply knowledge to some practical end.

Making the distinction between biology as a basic, "pure" science and technology as an applied science is not intended as a value judgment. Rather, it is essential to recognize that these approaches differ radically in their design, just as scientists differ in temperament and in their ability to pursue one scientific direction or another. Pursuit of curiosity about some aspect of nature cannot be charted in advance. The course of research is usually slow, erratic, and even fanciful. Discoveries often emerge from serendipitous encounters. Neither the tax-paying public nor the entrepreneur has much enthusiasm for investing in meanderings without goals or timetables. In pharmaceutical companies, in which it is standard practice to set "milestones" to measure progress toward a goal, meeting the milestones in a path laid out toward the development of a drug or vaccine is rewarded even when attainment of the goal is of rather modest difficulty.

Counterintuitive though it may seem to the layman (and, perhaps, even to the scientist), the most cost-effective advances in medicine are not likely to be made by frontal assaults on targeted diseases. Crusades against cancer have failed repeatedly, as have campaigns waged against other serious, complex afflictions. Wars are not won with words. When we are defeated, it is because we lack the weapons—the detailed knowledge of the chemistry and the functioning of cells that we need to fight disease. As a game, medical research resembles pool more than billiards: points are scored no matter which pocket the ball goes into, because each increment in technique and insight can benefit the efforts of researchers working on many different diseases.

Curiosity about how cells make the nucleotide building blocks of the coenzymes, RNA, and DNA led to revelations of the chemistry of the series of enzymes that make them from simple nutrients—carbon dioxide, ammonia, and amino acids. This proved to be the very knowledge needed to design the drugs now used to interrupt these synthetic processes in a variety of diseases—cancer, AIDS, herpes, and various autoimmune diseases (including rheumatoid arthritis, diabetes, and lupus)—and in the rejection of tissue grafts and organ transplants. In the same fortuitous vein, success in growing monkey kidney cells in test-tube cultures made engineering of the polio vaccine possible, and basic studies of the immune system led to the powerful technique for producing monoclonal antibodies.

Even when rational drug design fails, benefits may still accrue. Scientists were pleased when allopurinol, which they had synthesized for cancer therapy, proved effective against gout in a test patient; the drug emerged as an outstanding treatment for gouty arthritis. Nor were they embarrassed when acyclovir, which also had been intended as a treatment for cancer, turned out to be one of the best drugs against herpes infections.

The discovery of DNA polymerases and related enzymes that replicate, break, and repair DNA formed the basis of recombinant DNA and the breakthrough of genetic engineering. The revolutionary impact of this technology on medicine, agriculture, industry, and forensics has been justly heralded, but unmeasured and ultimately of even greater significance will be the insights gained into basic biological processes made possible by these techniques—

a deeper understanding of mutagenesis and carcinogenesis, of growth and development, and of disease and aging.

All these basic advances were made in academic laboratories built and supported almost entirely by funds from the NIH (National Institutes of Health). For thirty years, my research on the biosynthesis of nucleotides and DNA replication and the training of more than a hundred young scientists was funded with many millions of dollars without any promise or expectation of marketable products or procedures. No industrial organization would have had the resources (much less the disposition) to invest in such long-range, apparently impractical programs.

Faced with stringent budgets and the drive to provide universal health care at an affordable cost, the NIH will be under increasing pressure to focus on specific objectives and goals—AIDS, breast cancer, women's health—and to do less to support untargeted explorations of cellular chemistry and biology. Who will sponsor decade-long programs with no obvious applications that (for the time being, at least) merely satisfy scientific curiosity? No wonder that, in 1992, Dr. Bernadine Healy, then the director of the NIH, developed a "Strategic Plan for Medical Research." One of the questionable goals of the plan was "to expand the knowledge base in biomedical and behavioral research in order to enhance the Nation's economic competitiveness and ensure a continued high return on the public investment in research."

I am concerned that plans, even those with laudable goals, are fundamentally flawed, simply because they have not worked in the past. Discoveries are so commonly serendipitous that the best plan would seem to be no plan. For lack of essential knowledge, timetables for assaults on particular disease targets have little meaning. *Targets* and *focus* are the current buzz words. But targets and focus must continually change. No one could have anticipated our recent confrontations with such novel diseases as AIDS, Legionnaire's disease, and toxic shock syndrome nor the timing and scale of emergence of drug-resistant tuberculosis. Were it not for the advanced techniques of molecular biology and genetic engineering discovered in basic research, the dissection of the AIDS virus, refined diagnostic tests, and rational approaches to combat AIDS with vaccines, drugs, and novel genetic maneuvers would have been delayed by a decade or more.

A plan, by its very nature, cannot anticipate the utterly novel approaches that make possible major transformations in the acquisition and application of knowledge. The extraordinary success of the NIH, which has changed the face of medicine in the post–World War II period, was not planned. The response to the congressional mandate to establish and to fund institutes for twenty or more specific diseases came from Dr. James A. Shannon, who was the director of the NIH from 1955 to 1968, an interval in which the budget rose from $81 million to $1.5 billion. We are indebted to his wise leadership and to the assistance of two patron saints in Congress—Representative John E. Fogarty, a bricklayer from Providence, Rhode Island, and Senator J. Lester Hill of Alabama. They managed to channel a major fraction of the budgets of the heart, cancer, and other constituent institutes of the NIH into noncategorical basic research. Had this money been spent instead in palliating various specific diseases, the current promise of preventing and curing them would have been squandered.

In contrast to the "Strategic Plan" are the sober approaches proposed by Dr. Harold E. Varmus, who became the director of the NIH in 1993. In addressing the Conference to Establish a National Plan on Breast Cancer, he cited his extensive studies on the virus that causes mammary cancer in mice, a line of research that I think he chose because breast cancer had killed both his mother and his maternal grandmother. Varmus's studies on breast cancer led to the discovery of cellular oncogenes, a discovery honored by a Nobel Prize in 1989, which he shared with J. Michael Bishop. Actually, these studies yielded no insights into human breast cancer, but they did provide an important advance in understanding brain development.

At the same time, a research program headed by Dr. Robert Weinberg at the Massachusetts Institute of Technology (MIT) directed toward brain tumors in rats did make a major contribution to our understanding human breast cancer. Of the several genes now known to be involved in human breast cancer, all but one were discovered by working on something other than breast cancer. Clearly, the ubiquity of biological design means that a specific disease model as a target is far too narrow to encompass the complexities of a disease process, such as AIDS or cancer. Stated again, medical research is still a game of pool: you score no matter which pocket the ball goes into.

Another feature of Dr. Healy's "Strategic Plan" was to link the NIH to the economy of the United States. NIH programs would be designed to keep the U.S. pharmaceutical and biotechnology industries competitive in global markets in order to reduce health-care costs. These designs, interwoven as they necessarily were with economic, social, and political elements, would distort the missions in which the NIH, freed from partisan politics, had been (and can remain) supremely successful. The great strength of the NIH has been to foster the creative energies of able and motivated scientists. In so doing, it has made the most practical use of scientific talent and resources and has achieved cost-effective contributions of science to human welfare.

As important as the creation of new knowledge are the effective communication of such knowledge and its practical applications. Whether generated in academia or in industry, it is imperative that new knowledge be shared promptly through publication and that it be applied for the development of drugs, devices, and medical practices. Means should be sought to accelerate applications with due regard for an equitable distribution of rewards to the discoverers and their institutions and to the entrepreneurs for assuming risk and providing skill in the development of products. The hazards of secrecy and the abuse of patents need to be confronted, and these will be considered in Chapter 8.

The Two Cultures: Chemistry versus Biology in Academia and in the Pharmaceutical Industry

Chemists and biologists, with few exceptions, recognize that a gulf separates the practice of their two disciplines. While this gulf is not as wide as the one between the humanities and sciences to which C. P. Snow called attention (in his 1959 Rede Lecture at Cambridge, "The Two Cultures and the Scientific Revolution"), the fields of chemistry and biology constitute two distinctive cultures.

The historical roots of chemistry and biology are intertwined from early in the last century, and so are the conflicts between them—conflicts that go back to the polemics between Justus von Liebig and Louis Pasteur (regarding the role of the yeast cell in alcoholic fermentation) and that persist to this very day over research styles and directions. One might have expected that the coalescence of chemical and biological techniques that created genetic chemistry would have united these disciplines, but, paradoxically, it seems to be moving them even farther apart. For the chemist, the chemistry of biological systems is either too mundane or too complex. For the biologist, the intricacies of organic synthesis and the rigor of physical chemistry are beyond reach and irrelevant.

Chemists seek precise answers to well-defined problems, whereas biologists are content with approximate answers to complex problems. Chemists tend to be more conservative than biologists, even in their social and political attitudes. This lack of rapport is unfortunate, because biologists must look to chemistry for the foundations and future of biological science. While biologists are aware that enzymes determine the shape, function, and fate of cells and organisms, they shudder at the multiplicity and chemical complexity of enzymes and try to avoid thinking about them. As for chemists, three billion years of chemical experimentation by the biological world are not to be ignored but, rather, are to be appreciated for the limitless opportunities that are provided for exploring the awesome chemistry of nature.

Coalescence of the numerous basic medical sciences into a single, unified discipline is providing a more fundamental understanding of nature and will inevitably lead to even more remarkable—and *unexpected*—practical applications. This unified discipline of biological science has emerged because it is expressed in a single universal language: the language of chemistry. Much of life can be understood in rational terms if it is expressed in the language of chemistry. It is an international language, a language without dialects, a language for all of time, and a language that explains where we came from, what we are, and where the physical world will allow us to go. A language of great beauty, chemistry links the physical and the biological sciences.

Historically, the pharmaceutical industry has been dominated by chemists who synthesize an enormous variety of compounds for screening by pharmacologists in dozens of assays with the aim of discovering new drugs. Aside from insulin and a few small peptides, only small, orally ingestible molecules have qualified as drugs. The huge research budgets for pharmaceutical research had, until recently, avoided large molecules, such as the potent cytokines and antibodies made available by genetic engineering.

The tide is shifting. Biotechnology, either acquired from biotech ventures or developed in-house, is now increasingly supported by the major pharmaceutical companies. Biopharmaceuticals have become the focus for a biotechnology-driven industry.

Innovations will recognize the potential, for the prevention or cure of disease, of using the body's own agents—hormones, cytokines, antibodies, nucleic acids, polysaccharides—to promote or to antagonize the bodily processes in which those agents operate.

In academia, the gulf between chemistry and biology has, paradoxically, widened, owing to the power of the new techniques of genetic engineering. The chemist is attracted to the ever more refined examination of molecules and mechanisms, aided by designed alteration of proteins by specific mutations of DNA, by creation of antibodies with catalytic activities of enzymes, and by computer graphics of structures and functions. The biologist is emboldened to tackle the exceedingly complex problems of human development, aging, and degenerative diseases. The greatest promise of blending the two cultures is now to be found in industry, where projects demand a team effort between chemists and biologists and where success depends on developing truly novel drugs that are both safe and effective. In this setting, it is also possible to avoid the formidable barriers found in academia—departmental autonomy, traditional curricula, snobbery, competition for students and status, and an encrusted bureaucracy.

Although the schism between chemistry and biology is serious, it is far less a problem than the rising tide of public fear, distrust, and rejection of science, both chemical and biological. Chemistry has had a poor image for some time. "Better things for better living . . . through chemistry" had been the Du Pont Company slogan for many years. The slogan informed the public of the value of plastics, herbicides, and industrial chemicals for our individual and collective well-being. Then the logo was abbreviated to "Better things for better living." The phrase "through chemistry" was dropped when the public became aware that chemicals (as is true of all things, natural or synthetic) can be toxic. In fact, the only times we hear something good said of chemistry these days are figurative references, as in newspaper articles, to the "good chemistry" of a winning football team or to the "improved chemistry" between certain heads of state.

The image of the biologist has not fared well either. Hollywood has chosen biologists as their newest villains. Lacking communists as culprits, the makers of recent hit movies, such as *Lorenzo's Oil*, *The Fugitive*, and *Jurassic Park*, have demonized doctors and scientists. Never mind that well-controlled studies showed that Lorenzo's oil is without value, that criminal activity by a major drug company, as in *The Fugitive*, is exceedingly uncommon, and that the cloning of a whole dinosaur genome, as in *Jurassic Park*, is utter fantasy.

Perhaps Hollywood has taken its cue from congressional committees and the headlines they generate, which make it seem that science is wracked with fraud. Members of Congress and reporters both fail to recognize that the practice of science defines rather strict boundaries for behavior that are effective in all but the very rare instances of irrationality and criminality. In the practice of science, the more startling the claimed result, the more it attracts attention, and, if the claim is false, the sooner it is exposed.

Scientists are obligated to make it clear to the public that, although science is great, scientists are still people. As people, they are no different from others: doctors, lawyers, artists, merchants. Scientists are just as likely to fall prey to the human failings of arrogance, greed, fear, dishonesty, and psychopathy. What does set them apart from others is the discipline of science, a practice that demands exact and objective descriptions of progress, evidence that can be verified or denied by others.

It is the discipline of science that enables ordinary people, whether chemists or biologists, to go about doing ordinary things, which, when assembled, reveal the extraordinary intricacies and awesome beauties of nature. Science not only permits its practitioners to contribute to grand enterprises, it also offers them changing and endless frontiers for exploration.

Some may wonder whether the computer revolution and other advanced technologies have altered the way bioscience research is done these days. Can research now be engineered and pursued by formula? Not yet. The technical tools are indispensable, but the practice of science remains essentially an art form, and its province is nature.

Scientists probe the inexhaustible mysteries of nature from a variety of directions, and with different intensities and styles. These probings are determined by emotions, moods, and cultural heritage, much as these also influence the artist. The major discoveries in science are more often intuitive or serendipitous than the result of logical analysis.

The machines used by scientists produce images and compositions of objective precision. But this should not obscure the fact that scientists use these machines as tools, with personal tastes and styles as distinctive as those of painters in their use of color, as composers in their use of sound, and as writers in their use of words in creating their images of nature. Seneca, the great Roman statesman and philosopher, once said, "All art is but imitation of nature." What we try to do in science is to get ever closer to the secrets of nature.

In my own path toward understanding nature, I took a circuitous route that lately included excursions to biotech ventures. Starting with medicine and clinical research, the path veered to rat nutrition and then stayed on course for forty years in studies of enzymes, particularly those that make and manipulate DNA. The power of these enzymes to engineer genes and to produce massive quantities of precious hormones, cytokines, enzymes, and vaccines has alerted me to their potential value in the diagnosis, prevention, and treatment of a vast array of diseases. How biotech ventures have provided the most direct and effective means for application of these innovations is the subject of this narrative.

CHAPTER 2

Birth of a Biotech Venture

My scientific autobiography, *For the Love of Enzymes: The Odyssey of a Biochemist* (entitled in the Japanese translation, for no obvious reason, as *It All Began with a Failure*), describes my life in academic science. To a brief version of my autobiography included here below, I will add details of my burgeoning extracurricular activity in biotech ventures, particularly the founding and success of DNAX, which inspired me to write this book.

My Path to Biotech Ventures

Back in grade school in Brooklyn, where I was an eager and able student, it was customary, in the 1920s, to be advanced by "skipping" grades. After I skipped two grades and was being considered for a third skip, my brother Martin, who was 13 years older and a teacher then in the New York City schools, called a halt to my grade skipping out of concern for my social adjustment. At Abraham Lincoln High School, I made up for that lapse by finishing in three years instead of four, and I entered City College in 1933 at age 15.

What made Artie run? What made twenty-four students born of poor, immigrant Jewish families and graduated from New York City public high schools in the second quarter of this century become winners of the Nobel Prize, a record unmatched by any social group in the history of this award? To these laureates can be easily added an equal number, just as distinguished in science, who did not receive the Nobel Prize, and many more in the arts, business, and law. Nurture can drive people to achievement, and, once they are driven, their momentum is maintained.

The author

My parents and my sister Ella, nine years my senior, recalled many incidents of my restlessness and abandon as a child, pulling away from them and getting lost. This wanderlust was also evident in my choice of grade and high schools well outside my Bath Beach neighborhood in Brooklyn and in my willingness to endure a 3- to 4-hour daily commute by subway in order to attend the renowned City College in uptown Manhattan. Finally, despite my acceptance by a medical school in Brooklyn, I chose faraway Rochester, in upstate New York.

Why science? I don't remember having any interest or curiosity about my natural surroundings, except for collecting matchbook covers, the most abundant local flora of my youth. From age 9, I helped out in my parents' house-furnishings and hardware store, a feeble and unprofitable business, and then clerked in other stores, particularly in men's furnishings (haberdashery) shops, during evenings and weekends throughout high school and college and for summers thereafter. Unlike Paul Berg and Charles Yanofsky (close friends and cofounders of DNAX), I had no awareness of science research, either in high school or in college. Because I did well in chemistry courses and liked them, I had considered an academic career in the subject, but the employment opportunities in the mid-1930s were so dismal that I elected a premedical curriculum, with the agreeable prospect of four more years as a student in an M.D. program.

I enjoyed medical school, and I anticipated becoming an internist with academic connections. The courses in anatomy and physiology presented fascinating images of the intricate organization and functions of the human body; in bacteriology and pathology, we became acquainted with the morbid aberrations of horrid diseases. In classic medical-student fashion, I immediately succumbed to each of them—amyotrophic lateral sclerosis, aortic aneurysym, lymphoma—only to be cured by preoccupations with still other diseases. By contrast, biochemistry, with its boring enumeration of body constituents and their analysis in the blood and urine, offered neither a grand plan nor fundamental principles.

In the clinical years, including an internship in internal medicine at Rochester's Strong Memorial, the University Hospital (1941–1942), I was engrossed by the responsibility for coping with the complex problems of sick people. I felt compassion for and a deep rapport with distressed patients, but, despite an occasional diagnostic success, the lack of logical rigor and rational methods was discouraging to me. Signs and symptoms had to be promptly assigned a categorical niche, followed immediately by a prescribed treatment. There were few opportunities to think about the physiological basis of any disease. In these preantibiotic days, a fourth of all pneumonia patients died; service on a busy ward during the winter pneumonia season earned me a cash prize for obtaining the most autopsy permissions.

Medical students at the University of Rochester were encouraged by fellowship grants to do research. I applied for and was eligible for several, but, even with my superior academic record, I failed to get any. In some instances, the virus of antisemitism was surely at work. Still, I did collect data—at a borrowed laboratory bench late at night—to pursue my curiosity about my low-grade jaundice and the occurrence of the same symptom among other apparently healthy young people. Twenty years later, this benign deficiency in clearing bilirubin from the bloodstream, which produced a low-grade, chronic jaundice, was rediscovered as the inborn error of metabolism called Gilbert's disease, named after the Frenchman who first described it at the turn of the century. Publication of these data in 1942 in the *Journal of Clinical Investigation* caught the attention of Dr. Rolla Dyer, then the director of the National Institute of Health (NIH) in Bethesda, and was instrumental in my being transferred from sea duty in the Navy to do research in the Nutrition Section.

As a uniformed medical officer in the U.S. Public Health Service, I expected that my stay at the NIH would be a two-year tour of duty to be followed, most likely, by a clinical assignment. Instead, I found full-time laboratory research more fulfilling than any previous medical experience. I fed rats defined diets to determine their need for vitamins and minerals under various circumstances, a programmed exercise that could yield clear answers, unlike the jumble of problems confronted in the haphazard mix of patients who appeared in the hospital. Even though I had entered the field of nutrition in its twilight, I was still excited about the experiments I could design and the data I could collect, organize, and report. After a year or so, I thought no longer about a career in clinical medicine but, rather, about how to do more and better things in research.

The serious economic disadvantages of choosing a career in research over a lucrative practice of medicine really never entered my mind—a strange indifference, considering the poverty of my childhood and the many years of working, skimping, and borrowing to get through school. Nor were the consequences of this choice considered later when housing and maintaining a family made increasing demands on a modest, fixed salary. In the 1940s, academic positions were rare, research grants were nonexistent, and opportunities to do science were severely limited. What a vast difference from the landscape in succeeding decades, when research became a respected, well-compensated profession, and issues of geographic location, tenure, and retirement benefits were the main criteria for choosing among many attractive, alternative positions.

My work in nutrition contributed in a small way to the isolation of folic acid, one of the B vitamins, and it demonstrated that animals need it to make blood cells. But the way in which folic acid might be used by the body was a biochemical question phrased in language that was utterly foreign to me. My biochemistry courses in medical school gave scant attention to the emerging excitement about enzymes and their coenzymes, in which vitamins helped perform the catalytic functions. I became vaguely aware of revelations that the chemical energy stored in food could be converted to a common currency, called ATP (adenosine triphosphate), that made muscles contract and cells grow. I had also heard Edward Tatum describe in a seminar how he and George Beadle had used *Neurospora crassa*, a bread mold, to show that a single gene was the source of a distinctive enzyme. Still, the halls of the Nutrition Section did not reverberate with the thunderclaps of the emerging biochemistry.

At the conclusion of World War II, I persuaded William Henry Sebrell, my superior officer, to have the NIH sponsor my training in enzymology. It was clear by then that the vitamin hunters had about exhausted their prey and that the enzyme hunters would soon command the field. I stretched a few months' leave to a full year (1946) working with Severo Ochoa at the New York University Medical School, and I added a half year more with Carl and Gerty Cori at the Washington University School of Medicine in St. Louis. Purifying an enzyme from tissues, using the new spectrophotometric assay methods, yielded data in a matter of hours—an electrifying leap, compared with the months of waiting required to obtain a nutritional result. The capacity of an enzyme molecule to convert a thousand molecules of a substance to a product in a second with unfailing precision is a chemical feat that I still regard with awe and fascination.

Upon returning to the NIH in 1947, I organized an Enzyme Section with my close friends Leon A. Heppel, a classmate from medical school, and Bernard L. Horecker, a trained enzymologist who had introduced me to the rudiments of enzyme work.

With Herbert Tabor, who joined us in daily lunchtime seminars, we covered every facet of biochemistry in an academic atmosphere of doing and learning science, immune to the stigma of employment in a government laboratory, which so worried the Coris about my scientific future.

Those years of working at the bench all day, with Bill Pricer, my technical assistant, at my side, were golden years, in which I discovered how coenzymes are made and how inorganic pyrophosphate is generated, and in which I began my studies of how cells fabricate the nucleotide building blocks for assembly into RNA and DNA. Beyond the laboratory, the unwavering support of my wife, Sylvy, in raising our three sons and her deep understanding of science enhanced my capacity for this sustained research effort.

On the other hand, some of the concerns of the Coris were coming to pass, and I began to experience mounting irritation over the cumbersome and uninspiring NIH bureaucracy. This—coupled with a flattering invitation, in 1952, to return to Washington University as a professor and chairman of the Department of Bacteriology (renamed, then, Microbiology)—convinced me to move to a university setting. I had not anticipated the burden of teaching an unfamiliar subject, the turmoil of rejuvenating decrepit facilities, and the distractions of a different bureaucracy. These disappointments were, in large measure, offset by the discoveries of nucleotide synthesis and DNA synthesis and the assembly of a staff who would move with me, in 1959, to initiate the Department of Biochemistry in the Stanford Medical School, which was itself transplanted that year from San Francisco to Palo Alto.

With me came virtually the entire Washington University microbiology faculty, including Paul Berg and Robert Lehman, both of whom had been postdoctoral fellows with me, and Dale Kaiser, David Hogness, and Melvin Cohn; Robert (Buzz) Baldwin joined us from the University of Wisconsin. (All but Cohn have remained close colleagues in the Biochemistry Department to this day.) Our mean age was 34. The communal operations and spirit of the department led the group to focus sharply on nucleic acids. The wide range of our backgrounds allowed us to take approaches that ran the gamut from genetics to physical chemistry. Despite its youth and small size, the department soon became the seat of the recombinant DNA discoveries and the training ground for a large fraction of the future leaders in the field. The department was also a major influence in the renaissance of biochemistry and genetics at

Stanford, having encouraged Charles Yanofsky to accept a place in the Department of Biological Sciences and Joshua Lederberg to come from Wisconsin to start the Department of Genetics in the Stanford Medical School.

My ambition from 1950 on had been to discover the enzymes that made the nucleic acid polymers, RNA and DNA. But first we had to determine their building blocks (that is, nucleotides) and to learn how cells made them. Delineation of the many stages in the enzymatic pathways by which the building blocks are pieced together was an important step in understanding the biosynthesis of RNA and DNA—and, incidentally, it provided the targets later used for the design of most drugs currently used for the treatment of cancers, autoimmune diseases, AIDS, and herpesvirus infections.

In 1955, I found an enzyme, extractable from the common intestinal bacterium *Escherichia coli*, with the astonishing capacity to replicate DNA from any microbial, plant, or animal source. The synthetic product possessed the double-helical structure proposed by James D. Watson and Francis H. C. Crick in their epochal paper published two years earlier. The enzyme, named DNA polymerase, was the basis for award of the Nobel Prize for Medicine or Physiology for 1959, which I shared with my former mentor, Severo Ochoa. The award was a great honor for me and my family, but, despite its cachet, it did not affect my research or alter my lifestyle.

Although we could make DNA in the test tube with the requisite chemical and physical features of natural DNA, we failed for twelve years to demonstrate that it also had the genetic and biologic potencies of hereditary material. When we finally succeeded, in 1967, in making infectious viral chromosomes, media headlines reporting "creation of life in a test tube" evoked exaggerated attention. None of us in science expected that, in a few years, our enzymes and others that make and rearrange DNA would become crucial reagents in the preparation of recombinant DNA and the foundation of the genetic-engineering revolution.

Founding and directing the Biochemistry Department at Stanford, with its unique style and achievements, have been a source of great satisfaction for me. I have also enjoyed the authorship of several monographs on DNA replication and guiding the training of many gifted graduate students and postdoctoral fellows. But all of these other roles—administrator, author, teacher—were based upon, and remained secondary to, my devotion to science and the fulfillment it brought me.

Until 1968, the idea of my being associated with an industrial enterprise was utterly unattractive; it offered so little scientific reward. I had been distressed by several visits to pharmaceutical companies. Typically, a team of five or ten pharmacologists and chemists was assigned a major disease category—cancer, arthritis, hypertension—and asked to develop patentable drugs to compete with those already in use. These scientists, once young and eager, had become gnomes grappling hopelessly with problems far beyond their reach.

Along came Alex Zaffaroni with an invitation to join the scientific advisory board of ALZA. Zaffaroni had been my gracious host on a lecture visit to Syntex in Mexico in 1961, arranged by Carl Djerassi, who had recently joined the chemistry faculty at Stanford. I was taken with the style and vision of the Syntex operation and with Zaffaroni's leadership. I could also encourage him in his planned move of Syntex research along with establishment of pharmaceutical development in the Stanford Industrial Park. Our friendship had grown over the years, and, when he confided his plans to leave Syntex and to found ALZA, a company that would be dedicated to developing innovative drug-delivery systems, I was most supportive. My high regard for his scientific insights and integrity, and the possibility that I might contribute to the enterprise from my experience with membrane biochemistry and intermediary metabolism, persuaded me to compromise my virginity with respect to corporate associations and to accept his invitation.

During the twelve years that I served on the advisory board, I gained far more than I was able to give. I learned about polymers, pharmacology, and diseases, and especially about the hurdles of drug development—quality control, regulatory approval, and marketing. Most of all, I was impressed by the intelligence, ingenuity, and integrity of the many people Alex had assembled and charged with achievement at each stage of the enterprise.

I resigned from the ALZA advisory board in 1980 in order to have that time for my next biotech venture, the founding of DNAX. In subsequent years, I would join the scientific advisory boards of other companies—Regeneron, Metrigen (now out of business), XOMA, and GalaGen—each with different objectives and styles, but all trying to serve science and society in useful ways. Regeneron, which is located in Tarrytown, New York, seeks to do for the nervous system what DNAX is doing for the immune and hematopoietic (blood-forming) systems—to discover the hormonal agents that direct the growth and maintenance of specialized cells in these tissues. Metrigen tried to improve DNA sequencing with a novel use of DNA polymerase, but the support of venture capitalists was inadequate to weather technical problems. XOMA, in Berkeley, on whose board of directors I also serve, develops antibodies and other agents to combat infections and autoimmune diseases. GalaGen, in Minneapolis, uses colostrum from immunized cows for the production of proteins that are effective in the prevention and treatment of infectious diseases.

When I was invited to join each of these ventures, I declined, pleading that I could offer little in entrepreneurial zeal or technical expertise. My main contribution has been to convey the conviction that first-class science and technology offer the best foundation for a successful venture, and that this is best achieved by attracting the best talent and by providing the resources and ambiance for its creative expression.

My associations with enterprises and people outside the traditional academic sphere have been personally gratifying and scientifically enlightening. I have found most business managers bright and reliable. Their integrity and humaneness sometimes exceed those qualities among the managers of university affairs with whom I am familiar. The science, dealing, as it must, with the biology, chemistry, and diseases of mice and humans, takes me on excursions far beyond the limited focus of my own research on microbial systems. My evangelism for reducing biological phenomena to molecular detail and my faith in the universality of biochemical mechanisms throughout nature have had some impact. These convictions, combined with my early training and continuing interest in clinical medicine, have provided me with confidence in grappling with the scientific problems and strategies of these diverse biotech ventures. In the conception, birth, and development of DNAX, shared with Alex Zaffaroni, Paul Berg, and Charles Yanofsky, I found the rare opportunity to set standards for the conduct of research, to enjoy the science in biotechnology, and to share in its medical applications.

Conception of DNAX

Alex Zaffaroni listened eagerly, in our occasional lunch chats, to discussions of advances in the enzymology of DNA replication, repair, and recombination. He heard, in 1967, that we could use DNA polymerase to copy any template faithfully, even to the extent of creating an infectious viral genome. By 1972, the means to splice the DNAs from different genomes was achieved by two groups in the Stanford Biochemistry Department. With these recombinant DNA techniques, Stanley Cohen and Herbert Boyer prepared plasmids as vehicles to introduce foreign genes into bacterial host cells, where they could be replicated and expressed.

Excited by this new knowledge, Alex immediately wondered whether it might be applied clinically to produce novel drugs and medical devices. Could this genetic chemistry and molecular biology be a means of obtaining hormones and other potent macromolecules that are made in the body but that are present in amounts too small to be isolated for clinical use? Could genetically engineered DNAs become the "silicon chips" of a new industry? While theoretically possible, the uncertainties of these applications and the efforts needed to test them made the chance seem rather remote, in my estimation; but only a few years later, their feasibility was demonstrated in several laboratories.

The prospects of producing human insulin, growth hormone, interferons, and vaccines on an industrial scale began to intrigue some venture capitalists, who then enlisted chemists and molecular biologists to establish enterprises directed to those ends. By 1978, Genentech and Biogen were out of the starting blocks, and others were getting ready. Two years later, a Harvard University group organized a biotech venture called the Genetics Institute and invited Alex to assume its direction. He was tempted; but he was troubled both by ambiguities about the university's share in the enterprise and by the commuting distance to Cambridge from his ALZA base in Palo Alto. Yet, with Ciba-Geigy's ownership of his ALZA Corporation frustrating the wide application of its drug-delivery technologies, Alex was intrigued by the new vistas that ventures in genetic engineering could provide. In particular, he had long been impressed by the powers of the immune system and the extraordinary specificity of antibodies. By combining the newly discovered monoclonal-antibody technique with genetic engineering, it might be possible to tailor antibodies to become drugs and industrial reagents with a wide variety of uses.

Although I had resisted involvement in the entrepreneurial side of science, as had Paul Berg and Charles Yanofsky, we were persuaded, out of institutional loyalty, to engage in discussions of a Stanford venture. We had been invited by Alan Michaels, who was highly experienced in membrane filtration technology, and Channing Robertson, from the Chemical Engineering Department, to join a bioengineering enterprise called Engenics. Its mission was to improve the mass culture of microorganisms and animal cells and to perfect the purification of genetically engineered products. In meetings with the venture capitalists designated to finance the company, Paul, Charley, and I, close friends and responsible for some of the basic discoveries that made genetic engineering possible, were put off by their cold indifference to science and their intense focus on turning a large profit in two to three years. When I consulted Alex about the plans for Engenics, he reinforced our uneasiness.

Some weeks later, in October 1980, Alex was visiting me at Stanford, asking, in turn, for my advice about the Harvard company. When I told him he could find congenial colleagues closer to home, he said, "That's what I've been waiting to hear from you for a long time." We sealed the birth of our new, unnamed enterprise with a handshake, and we agreed to ask Paul and Charley to join. The three of us felt comfortable with Alex, aware of his extensive experience in business and science and his deep understanding of the pace and vagaries of laboratory progress. Above all, he was someone we trusted.

We Stanford three—Paul, Charley, and I—brought our experience and outlook in science to complement Alex's entrepreneurial talents and vision. Paul was a discoverer of recombinant DNA, a leader in the enzymology and molecular biology of protein synthesis, and at the forefront of the genetic engineering of vectors powered by tumor-virus genes. Charley, equally proficient in biochemistry and genetics, had promptly adopted and advanced recombinant DNA techniques to clarify still further the mechanisms and regulation of gene expression. I had discovered and characterized many of the enzymes of DNA replication that later became essential reagents in gene splicing and other DNA manipulations.

Infancy

Alex, operating from his ALZA office, took on the business affairs and, with advice from the three of us, named the new company, recruited scientists and staff, and selected initial projects. Toward these objectives, Alex applied his leverage. Perhaps his greatest use of it, even better than his intuitive grasp of how to exploit scientific principles and technologies for practical ends, is a talent for getting people to dedicate themselves to his ventures. Despite his muted way of articulating his visions and plans, Alex manages to excite scientists and businessmen alike and to sustain their devotion.

Because it cost only five dollars to register a corporate name in California, virtually all the euphonious *bio-*, *gen-*, and *nucleo-* titles had been appropriated. DNAX (pronounced dee´nax), which incorporated DNA and the popular terminal X (as in Xerox), was still available. To emphasize the primacy of research to our venture, we gave it the full and grand title: The DNAX Research Institute of Molecular and Cellular Biology, Inc.

We had no trouble in settling on the principles that would guide our choice of people and projects. We would select the best young scientists we could find, offer them the resources to work diligently and creatively, and trust that, in time, their associations with one another would lead to common interests and the identification of communal goals. Initially, we would not compete with those ventures already well started in the cloning of a number of known factors, such as erythropoietin, which stimulates the growth of red blood cells, or the interferons, which combat viral infections and cancer. Instead, we would focus on a specific area of biology and medicine. Immunology held a strong appeal. We decided to exploit biotechnology by going after antibodies, which could be obtained in quantity by the recently developed monoclonal-antibody technique. We would hone and redesign them in order to solve medical problems and, perhaps, even to serve some industrial purposes. With advice from medical colleagues and with our experience in molecular biology and protein chemistry, we would select suitable medical targets for antibodies and direct the engineering to make them.

German scientists, pre-eminent in the early part of this century, cited four G's as the essence for success in experimental work: *Geld* (money for resources), *Geist* (spirit and motivation), *Geduld* (patience and perseverance), and *Glück* (luck). An immediate DNAX need was *Geld*. Alex took care of that. He obtained $4 million from several Swiss bankers who had profited from previous investments with him. He incorporated DNAX in the Isle of Jersey, one of the Channel Islands, a maneuver designed to save taxes on future earnings, as had been achieved by making Syntex a Panamanian company.

Alex also obtained access to ALZA technology. In exchange for the use of ALZA's therapeutic delivery devices for a five-year period, ALZA was awarded 20 percent of the DNAX stock. Although DNAX never did exercise its option to use ALZA technology, the arrangement helped with its initial footing and funding, and ALZA profited handsomely, at a crucial time, from the sale of its DNAX holdings.

The Laboratory

The laboratory, we agreed, must be close to Stanford. Alex identified a building, one of several that ALZA owned in the Stanford Industrial Park, a five-minute drive from the university, that had been used for pilot-plant chemical syntheses by a failed ALZA spin-off and was now largely abandoned. It had also been used as a workshop and storage space for the ALZA buildings in the area. My dismay on viewing the dilapidated and cavernous spaces was allayed by Alex's reassurance that a clever architect and a million dollars could convert the 10,000 square feet into attractive laboratories and offices. He was right.

Alex chose Ken Kornberg, my youngest son, to design the laboratories. The distinctive features were close attention to the functional needs of the scientists, maximal use of space for laboratory work, and an aesthetically pleasing ambience. Rather than the conventional, straight, dim hallways, passages were angular and brightened by clerestory windows. Recognizing that eager experimental scientists work long evenings and weekends, spending many more hours in the lab than at home, friendly touches included a generous use of color. The choice of lavender to paint the halls and rooms brought a loud wail from the scientists, echoed by Alex. Ken bought time by pleading that repainting would cause delays and excessive cost and could readily be done after the rest of the job was completed. When the labs were finally occupied and the equipment, shelves, and posters covered most of the wall space, the occasional splashes of color were most welcome; no one wanted the lavender replaced by institutional pea green, tan, grey, or off-white.

Scientific Staff

Scientific staffing of a biotech venture in 1981 was difficult. The uncertain shape and future of a new enterprise and the stigma attached to an industrial position (when compared with an academic appointment) were daunting obstacles. We could promise our recruits only that we would strive to make DNAX a haven for serious science, offering staff members the finest resources to explore at a frontier of research and the freedom to communicate and publish freely and promptly. They also had our assurance of sustained devotion to the success of the venture.

The scientists who joined DNAX in 1981 had recently completed their postdoctoral training and were in line for academic positions in university departments. They saw, in DNAX, a favorable setting with access to technical assistance, supplies, and leading-edge equipment with which to make their mark in science. As they later learned, those resources also made for a more rapid ascent of the academic ladder by enabling them to skip several rungs. The financial inducements of equity in the company and a higher salary surely mattered, but they were mentioned only in passing. Recruited in this first wave were Kenichi and Naoko Arai, Bob Coffman, Frank Lee, Kevin Moore, Tim Mosmann, Gerard Zurawski, and Donna Rennick. A few words about each of them—their backgrounds and motivations—will introduce these young people, so diverse in personality, interests, and culture, and yet so amenable to applying their intelligence and ambitions to a communal effort.

Kenichi and Naoko Arai, husband and wife, had been postdoctoral fellows in my laboratory from 1977 to 1980, working on the enzymology of DNA replication in *Escherichia coli*. Their excellent Ph.D. training at the University of Tokyo was with Yoshito Kaziro, a world-class scientist and one of the foremost biochemists in Japan. They came to my laboratory directly from the airport, Kenichi with a clipboard listing more than twenty chores and detailed research plans (reminding me, in an offbeat way, of Bill Walsh, the celebrated coach of the San Francisco '49ers, who listed on his clipboard the first twenty-five offensive plays he would call in the game that day).

Kenichi Arai

Kenichi's early bent was to the physical sciences, but family traditions pushed him toward medical school. In preparing for the University of Tokyo entrance exams, a physics tutor acquainted him with papers in biophysics and molecular biology; the elective curriculum of the first two years, when he entered the university in 1960, gave him more time to become aware of the new excitement about DNA and genetics. Upon graduation with an

Kenichi Arai (right) *with his mentor, Yoshito Kaziro*

M.D. in 1967, political and social turmoil disrupted the university and forced a two-year interlude of part-time medical practice and readings in immunology and neuroscience. In 1969, he chose to do graduate research with Kaziro at the relatively independent Institute of Medical Sciences of the University of Tokyo (IMSUT). His Ph.D. thesis and subsequent work in a junior faculty position contributed to major discoveries of the mechanisms of protein synthesis and transmission of hormonal and cytokine signals by guanosine triphosphate–binding (GTP-binding) proteins. In 1977, he chose my lab for postdoctoral training because of an interest in the enzymology of DNA replication with a view toward the larger issues of how animal cells are induced to grow and proliferate. Kaziro and I had become acquainted two years earlier at a symposium in Madrid to celebrate the seventieth birthday of our mutual teacher, Severo Ochoa.

Kenichi is a dynamo. His energy and intellect are expressed in a remarkable capacity to master the basics and details of a wide gamut of biologic disciplines and in an ability to collect, galvanize, and propel the research efforts of a team of pupils and peers. His zeal in pursuit of a problem sometimes led him to intrude into the research of others, which caused trouble in my lab. I explained to those affected that Kenichi had no wish to appropriate their work but, rather, was impatient and driven to answer questions that had arisen in the path of his own research. My analysis was vindicated over the years when, on many occasions, Kenichi's personal and scientific devotion to colleagues inspired widespread trust and enduring friendships.

Naoko Arai

Naoko had attended the private girls' school in Tokyo associated with the Japanese Women's University, distinguished as one of the few college-level institutions open to women in pre–World War II Japan. Early in secondary school, inspired to rehabilitate polio victims and encouraged by her father, a practitioner in internal medicine, she decided to become a physician. Because the curriculum was designed to prepare girls for domestic life rather than academic careers, she had to educate herself for the rigorous entrance exams in math, physics, chemistry, and biology for the few highly contested places in the National Medical Schools. She was the only one in her class of 200 to apply to medical school. She accepted the offer from Yokohama University, but deferred a proposal of marriage from Kenichi, a marriage-related cousin whom she had met at family gatherings.

During the preparatory courses of the six-year medical program, she first heard about the recent startling advances in molecular biology and began to veer away from clinical medicine. Upon obtaining the M.D. degree in 1971, she entered the four-year Ph.D. program at the Institute of Medical Sciences to study under Professor Kaziro. Kenichi was already there, and the long incubation of his proposal culminated in their marriage. In 1975, their son was born, and Naoko remained in the Kaziro lab, first with a postdoctoral fellowship and then with a junior faculty appointment.

Naoko Arai

Kenichi's prodigious output during his three postdoctoral years—ten papers in one issue of the *Journal of Biological Chemistry*—attracted high-level attention. At Columbia University College of Physicians and Surgeons, Isidore Edelman, chairman of the Biochemistry Department, offered him a full professorship and Naoko a tenured associate professorship. At the University of Chicago, Nicholas Cozzarelli explored a position for them, as did Masayasu Nomura at the University of Wisconsin. Despite strong temptations to remain in the United States, they elected to return to Kaziro's group.

Back at the University of Tokyo, Naoko had no job, and she worked without salary or status. For Kenichi, the resources and independence of an assistant professor were woefully meager. Government support of facilities, training, and basic research, even at this most elite of the Japanese universities, had been, and was likely to remain, abysmal; in each minidepartment, the professor was the authoritarian director. Had Kaziro given Kenichi free rein to pursue new directions in DNA replication and the cell cycle in eukaryotes, it would have further diminished the group's capacity to focus their limited resources on GTP-binding proteins and their newly discovered connection with the *ras* oncogene.

Kenichi and Naoko were thus ready to accept my invitation to come to DNAX. They made this choice over the other alternatives because of their faith in our commitment to basic research and to provide resources that would support not only their own work but also that of promising junior scientists they wanted to bring with them from Japan, Atsushi Miyajima and Takashi Yokota among them.

It took courage for Kenichi to confront Professor Kaziro, their professional guardian, with their decision to join DNAX. Kaziro was stunned and upset. He wanted and needed Kenichi. To be abandoned by his most gifted student for an industrial enterprise was a disappointment and a personal affront. He could have tolerated their going to a well-known American university or even to the Roche Institute in New Jersey, but DNAX was beyond the pale. "I'll come to collect your bones" was all he could say. It was several years before Kenichi, vindicated by research success, returned to Japan and restored close relations with his mentor.

Robert (Bob) Coffman was poised, after immunology training at the Salk Institute and Stanford, to accept an assistant professorship at the University of Texas in San Antonio. He vacillated for several months before deciding to join DNAX in 1981. He was persuaded by Irving Weissman and other Stanford immunologists on the scientific advisory board of DNAX and by the charm of Alex Zaffaroni. The promise of a technician and postdoctoral fellow to do research of Coffman's own choosing would balance his involvement in the DNAX programs. He also saw fulfillment of a previously unsatisfied desire to apply research progress promptly to the solution of medical problems. As the first DNAX staff member near the scene, he served as a part-time consultant in ordering equipment and designing labs and as a greeter of new arrivals, generating and reflecting the excitement of creating a new enterprise.

Robert Coffman

Robert Coffman

Bob Coffman always wanted to be a scientist. From his home-town of Jeffersonville, Indiana, across the Ohio River from the big city—Louisville, Kentucky—he went to Indiana University in Bloomington. There he caught the new fever of molecular biology and was lucky enough, during his undergraduate years, to do research in molecular genetics in *Escherichia coli* under Arthur Koch. In 1969, he enrolled at the University of California at San Diego, where he came under the spell of Melvin Cohn at the Salk Institute. In a group of twenty or so bright

young people covering the gamut of cellular immunology, Coffman focused on a major issue—how immunoglobulin (antibody)
production was switched from the production of immunoglobulins of the most general class (IgM) to that of the more specialized classes (IgG, IgA, and IgE). In those days before recombinant DNA, he correctly concluded that the switch arose from
gene rearrangements and a process in which certain clones are
produced selectively.

Coffman's choice of postdoctoral training under Irving Weissman at Stanford broadened his perspective and introduced him
to the latest technologies of molecular and cellular immunology.
He also gained expertise in DNA manipulations, applying them
to studies of the influence of T cells on the switching of immunoglobulin classes and the exploration of the early lineage of B
cells.

Frank Lee continued to agonize about going to Columbia University in New York even after accepting an appointment there as
an assistant professor in the Biology Department in May 1981. It
worried him that, on revisiting the department, the professorial
staff did not seem to be happy in their work. He was also intrigued
by the sound of DNAX, by my conviction that the individual mattered more than the institutional connection, and by the promise
that he could continue his research interests. In November, with
the new laboratories ready, Lee changed his mind about the Columbia position and joined DNAX.

Frank Lee

Born in Susanville, California, in 1951 to parents from main-
land China, Frank could speak no English when he entered
grade school in San Francisco. The purchase of a television set
accelerated his learning of English (which, to this day, is not
used within the family circle). As an undergraduate at the University of California in Berkeley, he did research with Bruce
Ames on bacterial mutagenesis (later the basis for the Ames
test). He completed his Ph.D. with Yanofsky at Stanford on the
discovery of the control of gene expression by attenuation in
Escherichia coli. Frank pursued other aspects of regulation in

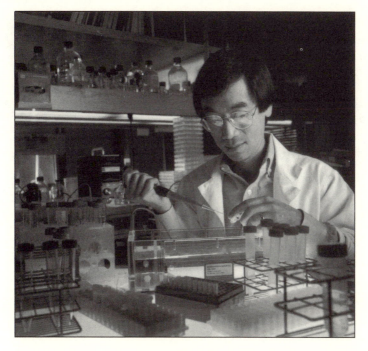

Frank Lee

postdoctoral fellowships with Philip Sharp at MIT, working on transcription of the adenovirus genome, and with Gordon Ringold, back at Stanford, on the control of gene action by steroid hormones.

Kevin Moore, trained in chemistry and exposed to immunology, was not enamored of the typical, successful academic career in which an early metamorphosis to managing and fund raising resulted in an estrangement from the laboratory bench. Made aware of DNAX by Leroy Hood, who had joined DNAX's scientific advisory board, he found the prospect of sustained hands-on work

Kevin Moore

and a communal sharing of research success very appealing. These features, and personal reasons for locating in the Bay Area, outweighed the risks of a new venture and made for a prompt decision to join DNAX.

Kevin Moore

Kevin grew up in New Jersey, entered Princeton as a National Merit Scholar, and graduated magna cum laude in chemistry in 1974, having done research in physical chemistry with Walter Kauzmann. His graduate work (in the Division of Chemistry at the California Institute of Technology, with John H. Richards) was on enzyme mechanisms, particularly those involving the vitamin B_{12} coenzyme. During a postdoctoral fellowship with Leroy Hood in the Division of Biology at Caltech, he became familiar with the new recombinant DNA technology in the context of rearrangements of immunoglobulin genes.

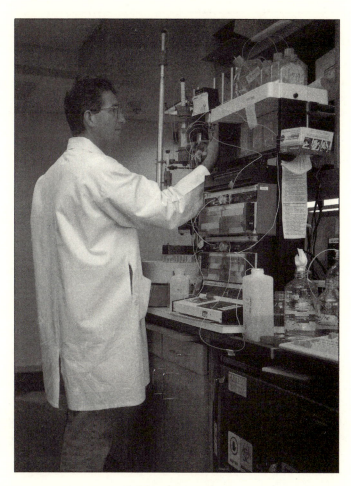

Gerard Zurawski

Gerard Zurawski, upon returning to Australia after postdoctoral training with Yanofsky at Stanford, lacked the sense he had felt in the United States of being at the cutting edge of science. Yanofsky's invitation to him to consider a position at DNAX promised new vistas and adventures in science and technology. With confidence in Yanofsky, and with the expectation he could always return to Australia or go elsewhere, if need be, he eagerly accepted the DNAX challenge.

Gerard Zurawski

Zurawski's parents met as refugees in postwar Germany and emigrated soon after to Australia. Gerard, born near Sydney in 1950, showed an interest in *Drosophila* (fruit-fly) genetics in high school, majored in microbiology in undergraduate studies at the University of Sydney, and worked on the molecular genetics of *Escherichia coli* for his thesis under Keith Brown. As a Fulbright Scholar and Fellow of the American Heart Association from 1977 to 1979 with Yanofsky, he followed Frank Lee in studies on the attenuation of gene expression and then returned to Australia to join the Plant Industry Division of the Commonwealth Scientific and Industrial Research Organisation (CSIRO) in Canberra to work on chloroplast DNA. Despite the security of a Queen Elizabeth Fellowship and the freedom of independent research, he chose to return to the United States to join DNAX.

Tim Mosmann left his faculty position at the University of Alberta to join DNAX for several reasons. Alex Zaffaroni had impressed him with the scientific and entrepreneurial directions of DNAX, as had the caliber of the scientists who had recently joined. He was also taken with the DNAX commitment to work on basic questions in immunology with the expectation that discoveries with the potential for medical products would be developed by pharmaceutical companies. An affinity for the West Coast of North America also weighed heavily in his decision to come to the Bay Area.

Tim Mosmann

Tim's roving, inquisitive mind matches the range of his rearing and scientific activities in four English-speaking countries. Born in England in 1949, he grew up in South Africa, where the discovery of the caterpillarlike velvet worm *Peripatus* in his garden enabled him to win a science fair—and may have inspired his own peripatetic inclinations.

At Rhodes University, he caught the excitement of the new microbiology and went on to graduate studies at the University of British Columbia, where he worked on the DNA of animal viruses. From there, he moved to Toronto for an introduction

Tim Mosmann

to immunology and protein chemistry. His discovery of mutant lines of B cells deficient in immunoglobulin secretion was extended by further studies during a Canadian MRC (Medical Research Council) Centennial Fellowship with Alan Williamson in Glasgow.

Later, as a faculty member in the MRC Immunoregulation Group at the University of Alberta in Edmonton, he worked on the production and use of monoclonal antibodies and on the analysis of the major histocompatibility complex. These studies attracted the attention of Irving Weissman while he was visiting the university, and Weissman then told Mosmann about DNAX.

Donna Rennick, after nearly twenty years of responding to her family's educational needs and movements, had completed graduate work in immunology and was fired with ambition to start her own career in science. DNAX gave her that chance.

Donna Rennick

Donna Rennick

Donna joined DNAX despite many impediments—the strong misgivings of her beloved and respected thesis advisor, the difficulty of transplanting her husband's established Sacramento medical practice, and the wails of three children forced to leave their cherished high-school friends. After earning her B.S. at the University of Colorado and eight years of experience as a medical technician, Donna chose the Medical Microbiology and Immunology Department at the University of California, Davis, when she entered graduate school, and she chose Eliezer Benjamini as her thesis advisor. Her work on anti-idiotypic antibodies was attuned to DNAX's project; she, in turn, was attracted to DNAX by the goals and philosophy expressed in interviews with several of DNAX's scientific advisors.

Atsushi Miyajima and Takashi Yokota, who had done outstanding graduate work under Kaziro (with immediate guidance from Kenichi Arai), were unhappy with their positions in Japan and eager to pursue the vigorous science they were promised at a postdoctoral level under Arai's direction at DNAX.

Atsushi Miyajima and Takashi Yokota

In graduate research, Miyajima had worked on aspects of transcriptional regulation, and Yokota had worked on factors in protein synthesis. In 1980, the Kaziro nest was overcrowded. Miyajima took an assistant professorship at Shizuoka University, where he had been an undergraduate; Yokota stayed with Kaziro as an assistant professor before assuming employment at the Toray Company in 1981, with the assignment to express beta interferon in *Escherichia coli* and to establish a molecular biology unit. Both Miyajima and Yokota grew impatient to get back to the vigorous basic research they had known. They could have waited a year or more for postdoctoral opportunities abroad, but they were attracted by Kenichi's offer of an immediate haven for doing science at DNAX and his continued benevolent concern for their future. The offer to Miyajima included the independence of working on yeast and a place for his wife, Ikuko, already well trained in biochemistry. Yokota was committed to Toray for a year and had to wait until early 1982. He also had the prospect of working on something new—signal transduction and DNA replication in eukaryotes—and a research assistantship for his wife, Kyoko. Kaziro tried to discourage them by emphasizing the uncertainties of an industrially sponsored laboratory in a strange land, but Miyajima and Yokota nonetheless opted to go with the Arais to DNAX.

William M. (Bill) Smith, a young patent lawyer with training in the biologic sciences, was an important member of the original staff. By keeping in daily contact with progress in all the research groups, he could help in writing reports and ensure that patents were filed promptly, without any delay in publication. A diligent attorney enlisting congenial cooperation of the scientists makes it possible to protect discoveries while avoiding the stifling secrecy that pervades most industrial-research operations.

Atsushi Miyajima

Takashi Yokota

Scientific Advisory Board

Alex Zaffaroni's approach to a new venture, as first applied at ALZA, is to assemble a board of scientific advisors, luminaries in diverse but related fields, who offer their expertise, provide a critical audience for reviews of the research, and, by their identification with the organization, assist in recruiting scientists and in raising the confidence of investors. The last of these functions might in some instances prove to be the most important, as Alex had learned from experiences with his ALZA boards.

In addition to Berg, Yanofsky, and me, the advisory board in the first year included thirteen others: William Dreyer, Avram Goldstein, Edgar Haber, Leroy Hood, Michael Hunkapiller, Kurt Isselbacher, Roger Kornberg, Thomas Kornberg, Ronald Levy, Harden McConnell, George Palade, Samuel Strober, and Irving Weissman. In 1982, Kimishige and Teruko Ishizaka were added to the board, and, with the Schering acquisition, Harvey Cantor joined as well.

Dreyer and Hunkapiller were associates of Hood, and they complemented his expertise in immunoglobulin genes and the sequencing and synthesis of polypeptides. Haber, Professor of Medicine at Harvard Medical School and Chief of Cardiology at the Massachusetts General Hospital (MGH), was well versed in antibody work and had produced the antibody selected as the first DNAX objective. Isselbacher, Chief of Gastroenterology at the MGH and Professor of Medicine at Harvard, was trained as a biochemist and had a broad knowledge of clinical problems. From the strong immunology community at Stanford, diffused as it was through several departments, Levy was focused on oncology and Strober on autoimmune diseases; Weissman, who had served as a member of Schering's task force that recommended its foray into immunology, was a leading figure in the basic mechanisms of immune modulation, a coauthor (with Hood and W. B. Wood) of the popular text *Immunology*, and very knowledgeable in clinical matters.

Also at Stanford, Goldstein, chairman of the Pharmacology Department, was an expert in drug actions; McConnell, a physical chemist in the Chemistry Department, was a pioneer in the theory and applications of electron spin resonance and active in biophysical approaches to immune phenomena; Roger Kornberg, in the Department of Structural Biology, was experienced in the struc-

Irving Weissman (left) *with George Palade*

tural features of membranes and chromosomes. Tom Kornberg, in
the Department of Biochemistry and Genetics at the nearby San
Francisco campus of the University of California, was active in the
burgeoning application of biochemistry and genetics to developmental biology. Palade, at Yale, honored with a Nobel Prize in Medicine or Physiology in 1974, was an anatomist who thought like a
biochemist and was steeped in virtually all aspects of experimental
biology and medicine. The Ishizakas, at Johns Hopkins University
Medical School, were the discoverers of IgE and were concerned
with its control in allergic states. Cantor, a Professor of Pathology
at the Harvard Medical School working at the Dana-Farber Cancer

George Palade (left) *with Yoshito Kaziro*

Institute, was a major influence in Schering's decision to target T-cell responses as an early research objective in immunology.

This large and congenial board, without any stated term of membership, evolved over the years. By 1992, of the original group, only Palade, Berg, Yanofsky, and I remained, joined by others whose expertise and personalities fit the changing needs and directions of DNAX. In addition to a modest equity and an annual retainer, the members were rewarded by their interactions with other board members, the excitement of the style and substance of the science at DNAX, and their share in the novel mission of linking basic research to the pharmaceutical industry.

The Science Plan

The training and orientation of the DNAX scientists and advisors strongly tilted toward biochemical approaches over those of cellular immunology, an imbalance that would be adjusted, and perhaps even overcorrected, over the next ten years. The plan was to identify the gene encoding a monoclonal antibody directed against some specific antigen. The gene would then be engineered to retain the part of the antibody that encodes the recognition and specificity-determining (Fab) fragment, which binds the antigen (Figure 1). At the same time, the section of the gene encoding the functional portion of the antibody (Fc), commonly the source of undesirable allergic responses, would be deleted. The gene for the Fab fragment would then be cloned and overexpressed in a bacterial host to produce kilogram quantities of the antibody for a medicinal need, or perhaps for use as a reagent at a particular stage in an industrial chemical process.

As a start and prototype for engineering antibodies, we were attracted to a project that had been conducted by Edgar Haber, a member of the DNAX scientific advisory board, whose knowledge of antibodies stemmed from postresidency research in protein chemistry with Christian B. Anfinsen at the National Institutes of Health from 1958 to 1961. They had demonstrated that an unraveled ribonuclease chain can fold back together into its unique three-dimensional structure, a proof that the intricate shape of the protein molecule is dictated by the singular sequence of the many hundreds of amino acids that make up the chain. This discovery earned Anfinsen the Nobel Prize in Chemistry for 1972.

Haber had been selected to be Chief of the Cardiology Division at the MGH by Walter Bauer, Chairman of Medicine. Haber's appointment was one of a succession of bold moves by Bauer, surely one of the most visionary figures in medicine of his generation. In building his department, Bauer chose a corps of bright, research-oriented young physicians, still clinically inexperienced, to be in charge of a specialty service staffed by a score of seasoned clinicians venerated for their practical and teaching skills. [Other selections of relative novices to direct major clinical divisions at the most prestigious hospital in the country were made at about the same time: Kurt Isselbacher for gastroenterology, Lloyd Holly Smith for endocrinology (later to become Chairman of Internal Medicine at the University of California, San Francisco), and Morton Swartz for

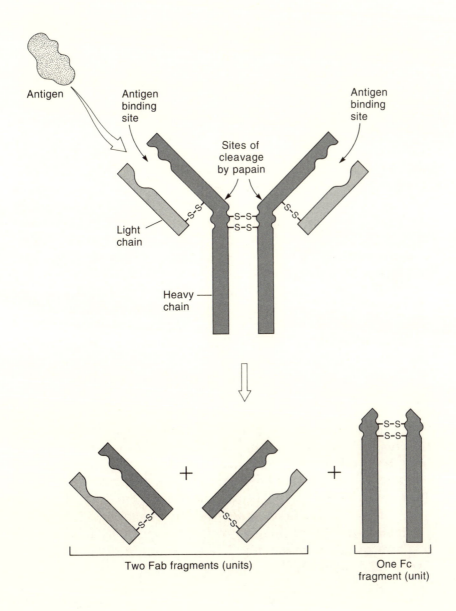

FIGURE 1. The Fab fragment of an antibody. An antibody is made up of a pair of heavy (long) chains (*shaded*) and a pair of light (short) chains (*unshaded*). Cleavage of the antibody by papain (a protein-cleaving enzyme from papaya) produces two Fab units (*F*ragment, *a*ntigen *b*inding) and one Fc unit (*F*ragment, *c*rystalizable).

infectious diseases.] These young recruits quickly acquired the clinical proficiency that earned them the respect of their older colleagues, while establishing vigorous research programs that advanced the knowledge base, practice, and teaching in their specialties.

At the MGH, Haber became interested in antibodies, and he showed that, as with ribonuclease, their refolding after denaturation restored their native structure and, with that, their antigen-binding specificity. Clinical experiences also made him aware of the problem of intoxication with digoxin (i.e., digitalis), which is widely used in the treatment of cardiac failure. Daily dosage of the drug and its long lifetime in the body result, not infrequently, in toxicity and even fatalities. Management of such emergencies entails watchful waiting for days while the drug is slowly eliminated. By administering the Fab fragment of sheep antibodies directed against digoxin to neutralize the circulating drug, Haber had been able to provide prompt relief from the toxicity and, in some dramatic instances, had saved people from virtually certain death.

In 1976, a preparation of sheep antibodies reversed the advanced intoxication of a 39-year-old man who had taken 22.5 milligrams of digoxin with suicidal intent; normal heart rhythm was restored ten minutes after infusion of 1100 milligrams of the Fab fragment. In Paris in 1979, a 34-year-old woman attempted suicide with 20 milligrams of digoxin and was nursed in an intensive-care unit for 43 hours until an air shipment of 817 milligrams of antibody promptly reversed the toxicity with no residual symptoms. The first use of such antibody preparations in a child, reported in 1982, promptly overcame the refractory ventricular fibrillations in a boy of two and one-half years who had swallowed an estimated 10 milligrams of digoxin. The ovine, digoxin-specific Fab, which was approved by the Food and Drug Administration for clinical use in 1986, is sold by Burroughs Wellcome under the trade name Digibind.

Stimulated by the DNAX mission to engineer antibodies, Haber set out to obtain highly specific antibodies by the *monoclonal* techniques discovered a few years earlier by César Milstein and Georges Köhler in Cambridge, England, a feat for which the two were awarded the Nobel Prize in Medicine or Physiology for 1984. Previously, the *polyclonal* mixture of antibodies obtained in the response of a sheep, mouse, rabbit, horse, or human to a foreign antigenic substance (such as a virus, a bacterium, or a chemical) included a large variety of forms. Each form of antibody in such a mixture is manufactured by one of the many different clones of cells produced by the immune system; each antibody is directed against a different facet of the antigen molecule.

Mixtures of polyclonal antibodies are limited in supply, and they have properties that vary greatly from one batch to another because they are produced by repeated immunization of individual animals. By contrast, a monoclonal antibody, produced by the clonal descendants of a single cell, isolated from an animal and maintained in culture, can be obtained as a homogeneous, highly uniform agent, propagated indefinitely, and produced time and again in the huge quantities needed for treatment of a disease. This is what makes monoclonal antibodies a preferred material for research and an attractive product for pharmaceutical use.

Based on Haber's favorable results with a polyclonal antibody, we expected that a monoclonal antidigoxin product would be sorely needed and would be stocked in hospitals worldwide. It seemed to us that the use of an antibody to detoxify a drug overdose could be extended to patients poisoned by other drugs and agents, such as narcotics, barbiturates, benzodiazepine sedatives, anticoagulants, and hormones, such as insulin. Medical reports indicated that significant numbers of distressed people could be saved if emergency rooms around the country had a battery of antidotes to neutralize toxic levels of these drugs.

The genetic and immunochemical means of preparing "designer antibodies," combined with the sophisticated delivery by ALZA polymeric devices, was novel enough to be referred to in a September 1981 notice in *Scientific American* as a "New Discipline of Immuno-Genetic Engineering." The engineered molecules would be designated as Minimum Binding Site (MBS) peptides for use in the diagnosis and treatment of diseases and in chemical processes.

Clinical applications were suggested, and trademark names were coined for developing MBS peptide products: as a detoxicant, Dig-Annul to scavenge overdoses of digitalis; as an imaging enhancer, Thrombo-Vis to visualize a blood clot; as therapeutic aids, Cytorad and Thrombo-Lys to direct drugs to specific tissues or vascular deposits; as drug-targeting agents, Tenin-Dart and Beta-Dart to deliver receptor antagonists or inhibitors; as an external extractor, MG-Extractor for myasthenia gravis; and, as industrial reagents in affinity separations, Selectomer resins for purification procedures.

License agreements were concluded with MGH for options to Haber's inventions and discoveries in his research on the MBS products, and also with Stanford for use of human monoclonal-antibody systems, should the exploratory studies of Henry Kaplan, Professor of Radiology at Stanford, prove successful.

The Business Plan

The average cost per scientist in a biotech venture in 1981 was, by rule of thumb, about $100,000 per year. With the remodeling of DNAX laboratory space costing $1,000,000 and the bills for equipment, supplies, and travel adding up to a like sum, the funds remaining of the initial $4,000,000 could sustain a staff of twenty for only one year. Alex operates in a generous style, but even skimping to reduce the "burn rate" could extend the life of DNAX for only a matter of months. We urgently needed more money, and 1981 was an especially lean year for biotech investments.

To get the next financing, we would try to persuade pharmaceutical companies and investors to take a stake in DNAX or to share in the research costs. In July and October of 1981, I accompanied Alex to a dozen or more meetings, in Japan, France, England, and the Bay Area, to plead DNAX's case.

A proper business plan had been prepared, largely with the assistance of William P. (Bill) O'Neill, who had worked for Alex at ALZA and was now enlisted at DNAX as Vice President for Corporate Development. Bill had obtained his Ph.D. in bio-organic chemistry at the University of Illinois in 1966 and, with a family to support, had taken a job at the Shell Laboratories in Emeryville, California (near Berkeley), working on biological membranes. After seven years, he joined ALZA. Curious about, and confused by, the apparently irrational and idiosyncratic business decisions at Shell Laboratories, a science-based company, he had hoped to develop some financial insight by pursuing an M.B.A. in evening courses at Golden Gate University in San Francisco.

Alex turned to O'Neill for advice when he had been invited in 1980 by a group of Harvard University people to direct the Genetics Institute, a genetic-engineering company that would follow the examples of Genentech, Amgen, Centocor, Immunex, Applied Biosystems, and Becton-Dickinson in the industrial development of leads provided by recombinant DNA and monoclonal antibodies. [That year, O'Neill had prepared for James Gibbons (then Director of the Congressional Office of Technology Assessment, and now Director of the Office of Science and Technology Policy in the White House), the landmark "Summary of Entrepreneurial Ventures in Biotechnology," a 250-page volume that is still in print.] After Alex had decided not to join the Genetics Institute but, instead, to found a biotech venture in the Stanford area, he was able to persuade O'Neill to join DNAX full time, and he gave him the charge to develop its principles into practical business programs.

Our first fund-raising foray was to engage Japanese drug companies. On our way to Japan, Alex, Ed Haber, and I stopped for a day in Honolulu to rehearse the slides and texts we would use in our presentations. I would begin by explaining recombinant DNA and genetic engineering, an area of science still unfamiliar to most pharmaceutical people at that time. I summarized Berg's new technique of cloning DNA, Yanofsky's new insights into the attenuation of gene expression, and my months-old discovery of an enzyme system for the initiation of chromosome replication. Haber would present his digoxin antibody studies and related work in the cardiovascular field. He would point out the physiologic advantages of the Fab fragment relative to the intact antibody: rapid distribution in the body and rapid excretion from it, the absence of troublesome aggregates of the antibody with body constituents, and the minimized risk of forming antibodies against the administered antibody. The enormous power of monoclonal antibodies over the conventional polyclonal preparations would also be emphasized.

To Alex was left the description of DNAX and its business plans for the future. In addition to the uses of the engineered antibodies to reverse drug overdosage and intoxication, he would point out their potential for reversal of an endogenous hormone, for imaging a unique antigen, and for the blockade of a receptor, and he emphasized, of course, how ALZA devices might be used in the targeted delivery of therapeutic agents.

At some point, we would all explain how we intended to develop a first-class research institute operating in an open academic atmosphere linked to pharmaceutical companies for product development of DNAX discoveries. We would: (1) attract the best young scientists, such as Kenichi Arai; (2) keep them by providing good facilities and full freedom to publish, to exchange ideas, and to focus on problems in depth; and (3) ensure the protection of intellectual property by having a patent lawyer with scientific knowledge (Bill Smith) as a key resident member of the research team.

We met with groups of six to ten people from each of several companies: Suntory, Takeda Chemical Industries, Wakunaga Pharmaceuticals, Fujizaki Pharmaceutical, and Yamaichi Securities. I knew the president or research director of the first three; we had introductions from Japanese friends to the last two. The receptions were cordial, the listeners were attentive, many photos were taken, and the meals were memorable (the French cuisine in Tokyo and Osaka was easily of three-star quality). However, we found no overt interest in investing in DNAX. I was keenly disappointed. When I complained to Alex that, in nearly thirty years of applying for grants, I had never been turned down, he said, "Get used to it. I was turned down more than forty times seeking help for ALZA." Thus did my internship in financing commence. I was learning that even powerful and exciting scientific ideas and approaches may lack the strength to open an investor's wallet.

In Paris, the late Claude Paoletti—once a postdoctoral fellow at Stanford and by this time well connected through his pharmacological research with pharmaceutical companies—opened the door to the Sanofi Corporation. We found a faint glimmer of interest there, but no cash.

In London, we were introduced to Victor Rothschild at a small dinner party given by his cousin Marcus Sieff, President of the Marks and Spencer stores. I had known Lord Sieff since 1962, when I joined the Board of Governors of the Weizmann Institute in Israel. Lord Rothschild graciously listened to our DNAX story, and he arranged for us to meet the next day with David Lethers at the Rothschild investment offices. We were turned down, presumably based on the negative advice of their scientific advisors, at least one of whom I expected to be sympathetic. Months later, in a second overture, Paul Berg and Alex met in New York with a Rothschild staffer and a young Rothschild heir, a molecular biologist with a Ph.D. from the Medical Research Council unit in Cambridge. After an animated and congenial scientific discussion over dinner at the famous restaurant La Cygne, Berg was convinced that a deal was in the bag. But conviviality can be deceiving; nothing was ever heard from them.

While in London, Alex and I had a most responsive audience in the offices of the *Economist*, meeting with the magazine's science reporters and their editor, Andrew Knight, whom I had met at Stanford. They did a fine story on DNAX in their characteristically incisive, colorful style, making me a faithful subscriber and an evangelical supporter of the magazine.

Meetings with potential investors in our own Bay Area ended in failures. To Standard Oil in San Francisco, our open research atmosphere seemed impractical. To George Shultz, who was then the president of the Bechtel Corporation, it was essential that Bechtel know in advance who the other investors would be: "Who will be in bed with us?" At that point, we had no bedmates.

Our most discouraging encounter was at Syntex with John Fried, President of Syntex Research, and a group of his scientists. They listened listlessly to Alex, Yanofsky, and me, and then left us without even a feigned murmur of interest. Not only did they lack any appreciation for what DNAX molecular biology and immunology could do for Syntex, they were utterly rude in their treatment of Alex, who had conceived and created the Syntex they occupied in Palo Alto. When I vented my feelings to Alex about the shabby way he had been received, he simply shrugged.

We tried to interest other American pharmaceutical companies in supporting DNAX. A dinner meeting with Roy Vagelos, President of Merck, was another disappointment. Vagelos had been an outstanding biochemist whom we all knew well and who held us in high personal and scientific regard; yet Vagelos revealed disdain for what DNAX and other small biotech ventures might achieve, and this disdain would dictate Merck policy for the next decade.

Alex continued the search for funding through the end of 1981, but I cannot recall that he expressed any anxiety, even when the company's money would soon run out. He was confident that something would happen, and it did. Our savior was Schering-Plough. What brought them to DNAX and ultimately to the marriage altar were the people on each side—Alex and his colleagues on ours, Bob Luciano and his associates on theirs. Congruence of ideas and plans was essential but hardly sufficient. What mattered was the mutual trust built upon past performances. The experience, talent, and character that Zaffaroni and Luciano brought to DNAX need to be recounted in order to acquaint the reader with the principles and practices of this new and unique venture.

CHAPTER 3

A Scientist–Entrepreneur: Alejandro Zaffaroni

Recipes for a humane and profitable business whose ingredients are innovative science skillfully blended with industrial applications are very rare, but such a recipe was possessed by Alejandro Zaffaroni, an uncommon native of Montevideo.

Youth in Uruguay

In a small Italian town, north of Milan, the flour mill could not support yet another Zaffaroni. Like millions of other Europeans in the nineteenth century, Alex's grandfather sought, in the New World, the better life described in letters from friends and relatives. He settled in Montevideo, Uruguay, at the age of sixteen. He worked hard, wed another Italian immigrant, and supported the large family into which Alex's father, Carlos, was born. By dint of ambition and effort, Carlos learned accounting in night school, became the manager of an alcohol fermentation plant, and finally the president of a bank, which, under his direction, became one of the largest in the small and thriving city. With Alex's mother, who was descended from an old Spanish family, five children were raised in the comfort and discipline of a devout, middle-class family.

Alex, the youngest, was born in 1923. He lost his mother when he was twelve years old, and his father died six years later. During that interval, he was his father's companion at concerts, for an occasional overnight boat trip to Buenos Aires for the opera, and in discussions of urban and national affairs, in which Carlos played advisory roles. Their bonding withstood Alex's indifferent performance at the primary and secondary Jesuit schools and his early departure from the faith.

His only inspirations from the strict Jesuit discipline and authoritarian rules were the schemes he devised to escape them. Instruction in science was virtually absent. Imagine the relief he first felt on breathing, at the age of seventeen, the open and permissive atmosphere of Montevideo University. Then imagine the shock of facing rigorous final examinations, ill-prepared in biology, chemistry, and physics. To Alberto Barcia, a fellow student and a close friend ever after, he feels indebted for helping him to learn how to study and for his grounding in the basics of chemistry.

Alex's entrepreneurial skills were apparent very early. In his second year at the university, he arranged with five other students to share the expensive books they needed, and, with the money saved, he rented a small apartment in which they studied and informed each other of scientific topics at weekly seminars, all leavened with fun—playing a Spanish card game called *tute*. Out of the discipline of this miniacademy, they emerged as more serious students and lifelong friends.

In the Faculty of Medicine, two years of basic courses preceded the six years of clinical studies. A compassionate desire to help people had directed him to medicine, but he failed to survive the trials of gross anatomy, histology, and embryology, all taught with obsessive attention to the minutest of details. For Alex, who to this day cannot find his way with a simple road map, the three-dimensional twistings and turnings of arteries and nerves in and out of tissues and organs were a bewildering maze.

Chemistry, by contrast, fired his imagination. He often skipped medical-school classes to attend chemistry lectures in the nearby Faculty of Science. The medical-school course in biochemistry was an exception. In what was regarded as the toughest medical course, Alex got the top grade and an invitation to be the teaching assistant. The professor, Carlos Amorin, a gifted teacher who gave enzymes and hormones special attention, rekindled Alex's childhood recollections of the magical displays of chemistry: the miraculous appearance of a blood-red solution upon mixing two colorless fluids, or the lighting of a tiny bulb connected by wires to copper and zinc strips dipped in beakers of acidified water.

Were Alex to adopt a middle name, it would probably be *Leverage*, a word he has used countless times to define a catalytic or nucleating event. What struck him early was the capacity of enzymes and hormones in minuscule amounts to alter profoundly the physiology and behavior of cells and organisms. Later, in his graduate research, he was impressed that the discovery of a simple technique to separate and analyze steroids would have profound consequences for a wide range of problems in chemistry and medicine.

Biochemistry is what he wanted to do; the rest of the medical curriculum bored him. But at the time, Uruguay, like Europe, offered no advanced training in biochemistry, aside from career paths in medicine and organic chemistry. In the library of the American embassy, Alex found catalogs that described Ph.D. programs in biochemistry in American medical schools. With the help of a friend with a better command of English, he applied to, and was accepted for, graduate work at Harvard Medical School, enabling him to get a student visa and priority for a one-way ticket on an overcrowded cargo vessel, one of the so-called "Victory Ships," bound for New York. It was July, 1945, and the war was still on.

The Harvard Medical School Biochemistry Department operated under the conventional system, in which the professors chose and directed the research of graduate students. Alex was disappointed by this lack of freedom. Also, with the department chairman absent in Washington on a war project, the research programs were in disarray.

Prospects seemed brighter and freer at the University of Rochester in upstate New York, to which he had been awarded a tuition scholarship. The chairman of the Biochemistry Department at Rochester, Walter Ray Bloor, absorbed in a war-research project, gave his students free reign, with a preference that they work on lipids, his focus of interest. To Alex, the opportunity to pick his project and have his very own lab was daunting but irresistible. So, to make his mark in science, he chose provincial Rochester over internationally famous Harvard.

Rochester, a small city of close-knit families and harsh winters, could have been a lonely place for a young, single student from Uruguay. Fortunately, Alex's hometown sweetheart, Lida, joined him a year later, but they first had to be married by proxy. Their first child, Alejandro A., now an ophthalmologist in San Francisco, brought them additional company in 1950; a second child, Elie, was born three years later in Mexico.

Graduate Study in Rochester

Although the Biochemistry Department at Rochester focused on phospholipids and the even less functional neutral fats, Alex was drawn, naively but strongly, to the life-essential hormonal steroids, particularly those produced by the adrenal gland. More than twenty steroids had already been extracted from the adrenal gland, but, for lack of analytic methods, the biosynthesis and fate of the physiologically important adrenal steroids remained unknown.

Although Alex had a lab bench, there was not enough money to buy reagents or simple equipment. Once again, his business sense kicked in. By collaborating with Henry Keutmann and Robert Burton in the Endocrinology Division of the Department of Medicine, he could get support to explore ways of measuring the adrenal steroids in blood and urine.

Alex's graduate work, done entirely on his own, set the pattern of his future career—an innovative use of physical methods and novel technologies to advance the understanding of basic life processes and to generate procedures and products of great practical value. He learned how to adapt the technique of separating compounds coursing through a sheet of paper (paper chromatography), recently discovered for water-soluble compounds, to the separation of the water-insoluble steroids. Intrigued by the various unique and vivid colors of the steroid spots when made visible through treatment with sulfuric acid, he devised sophisticated ultraviolet analyses of those spots, which were made possible by the newly available Beckman DU spectrophotometer. By applying these micromethods to the quantitative separation and identification of steroids, he provided access to the vast zoo of adrenal hormones and the means to track them wherever they might be.

Another feature of his future style, evident at this very early stage, was a consummate skill in arousing the enthusiastic involvement of a wide array of people in applications of his work. With Robert Burton, he examined the steroids in blood and urine. He collaborated with Oscar Hechter and Gregory Pincus of the Worcester Foundation for Experimental Biology in Shrewsbury, Massachusetts, to illuminate the biosynthetic pathway of corticosteroids from acetate and cholesterol. He helped Leo Samuels of the University of Utah to identify compound F, the circulating adrenal steroid in human blood, as cortisol (hydrocortisone). To apply his techniques to the mass screening of steroids in the urine of

Alejandro Zaffaroni as a graduate student at the University of Rochester

cancer patients, he enlisted oncologists at the Sloan-Kettering Institute for Cancer Research in New York. He also gained the attentive cooperation of the leading steroid chemists around the world: Tadeus Reichstein in Basel, Edward C. Kendall at the Mayo Clinic, and, most significantly, George Rosenkranz and Carl Djerassi at a small chemical company in Mexico known as Syntex, S.A.

It was Alex's good fortune that, in 1949, the year when he reported his novel techniques, Philip Hench and Edward Kendall also made the dramatic announcement that arthritis could be relieved by cortisone. Pictures of previously crippled people walking with ease after treatment astonished the medical world, captured newspaper headlines, earned them the Nobel Prize only a year later, and started a stampede among drug companies to meet the enormous demand for cortisone. D. H. Peterson of the Upjohn

Company, one of several who had come to Alex to learn his new methods, applied them to discover the microbial modification of progesterone to cortisone, a process that became a multimillion-dollar bonanza for Upjohn.

Alex applied his ability to analyze adrenal steroids to explore the capacity of cell-free extracts of adrenal glands to insert the essential chemical unit (a hydroxyl group at the 11-β position) into a variety of possible steroid precursors. He expanded the number of trials through improvised, automated, standardized extraction procedures, which he later expanded at Syntex for the thousands of trials that led to Synalar, a dermatological anti-inflammatory drug, the great maiden product of Syntex.

Alex's collaborations with Syntex would prove to be of the greatest importance. After obtaining his Ph.D. in 1949, Alex stayed on at Rochester for two more years, supported by a National Institutes of Health fellowship. He had intended to return to Uruguay, but there were neither the positions nor the climate for serious steroid research. Although academic appointments in the United States were scarce, he was offered an assistant professorship at the University of Rochester, a position at Harvard in Louis Fieser's group in the chemistry department, and a research position at the Worcester Foundation. However, a meeting with George Rosenkranz (then the research director at Syntex) at a Laurentian Hormone Conference in 1950 drew him into the Syntex net.

Alex was captivated by the intellect, warmth, and sensitivity of Rosenkranz. There was an instant rapport, and he readily accepted an offer to be a consultant to Syntex as a possible prelude to joining the company. On his first visit to Mexico, he met Carl Djerassi and was taken with his dynamism as well. In the ensuing years, the triumvirate of Rosenkranz, Djerassi, and Zaffaroni would find fame and fortune in science and business, would bring Syntex to its corporate success, and would knit their families together in fast and enduring friendships.

Before joining Syntex, Alex had to be assured of the independence he would have as Assistant Director of Biochemical Research. To attract him, Rosenkranz provided Alex wide latitude to pursue his own work on adrenal hormones in a free and open atmosphere and to communicate and publish his findings without restraint—freedoms uncommon in the pharmaceutical industry. An additional lure was the opportunity to teach bright University of Mexico graduates to become research scientists, and to do so in a language and setting that touched his Latin roots and sentiments.

Scientist–Entrepreneur at Syntex

The origins of Syntex go back to a maverick, ingenious chemist, Russel E. Marker, who encountered (or created) trouble wherever he went. In 1934, while at the Rockefeller University (then still the Rockefeller Institute), he wanted to improve the synthesis of steroid hormones. Progesterone, when available, cost $1000 per gram, and its use in female endocrine disorders was infrequent and usually insufficient. When the head of the Chemistry Department refused to let him work on steroids, he moved to Pennsylvania State University, where he discovered how to remove the bulky side chains from diosgenin, an abundant plant steroid, so that it could become a starting material for progesterone synthesis.

Mexican yams of the genus *Dioscorea* were said to be a rich source of diosgenin, with the roots of jungle plants weighing up to 100 kilograms. Marker made his way to Mexico in 1941 and, after harrowing misadventures, finally found enough material to determine the feasibility of producing progesterone from extracts of these roots. Back at Penn State, strapped for money and laboratory space, he still managed to produce an incredible 3 kilograms of crystalline progesterone.

In 1943, he set out again for Mexico to get financing for the on-site production of hormones, and, in 1944, after much effort, he convinced Emeric Somlo, the owner of Laboratorios Hormona, to start a new company. They called it Syntex, S.A. Within a year, Marker had produced 40 kilograms of steroid hormones, but, as he alleges in a memoir, was bilked out of the 40 percent equity in the company that he had been promised. So, in 1945, he set up a new laboratory in a small town outside of Mexico City, using the *Dioscorea* plant locally known as barbasco, a rich source of diosgenin, and, with the efforts of several young women who had worked with him at Syntex, again started making large quantities of steroids. By his account, the physical abuse of his workers by unknown criminals, lawsuits by Somlo, and the plummeting price of progesterone (from more than $80 per gram to less than 40 cents per gram) forced him into retirement from chemistry. The arrival of Rosenkranz in 1945 and of Djerassi four years later gave Syntex the impetus and research capability it had lost with Marker's departure.

Rosenkranz had escaped the Nazi terror in Hungary via Zurich, where he obtained his doctorate in chemical engineering under Leopold Ruzicka, renowned in steroid chemistry. He then went on to Cuba, where, in 1941, he found employment in a primitive pharmaceutical laboratory. Aware of Marker's publications and familiar with plant steroids, Rosenkranz had already managed to make some sex hormones in Cuba from the roots of Mexican sarsaparilla (*Smilax* species). In Rosenkranz, a fellow refugee, Somlo found the man he needed to direct the research at Syntex and to organize the large-scale production of hormones. Rosenkranz is a brilliant chemist, whose analytic mind has also allowed him to become an international grand master in contract bridge and the author of annual volumes on game strategy and anecdotes of tournament play. His scientific talent, broad interests, and infectious enthusiasm persuaded Carl Djerassi, in 1949, to leave Ciba (later Ciba-Geigy) and to join him at Syntex. At that time, Djerassi was 26 and Rosenkranz was 33.

Djerassi, born and raised in Vienna, had escaped the Holocaust in 1939 by emigrating to the United States by way of Bulgaria, where his paternal roots went back for centuries. In successive years, he attended a junior college in New Jersey, and then obtained a B.A., summa cum laude, from Kenyon College in Ohio. As a research chemist at Ciba in Summit, New Jersey, he shared in the discovery of Pyribenzamine, the first antihistaminic drug. After establishing a speed record by completing the work for a Ph.D. degree in only two years at the University of Wisconsin, where he caught the attention of William Johnson, Djerassi returned to Ciba. (Johnson, an outstanding organic chemist, would move to Stanford in 1960, revamp its chemistry department, and attract Djerassi in the first of many coups in recruiting a stellar faculty.) Djerassi's subsequent, gargantuan achievements in science and business and in the patronage of the arts are best learned from his delightfully informative autobiography, *The Pill, Pygmy Chimps, and Degas' Horse*.

The furiously paced activity of Rosenkranz and Djerassi, assisted by eager young chemists, produced a flood of publications. Their fine steroid research attracted wide attention and earned for Syntex the sobriquet of "University of Steroids" from the eminent Harvard chemist Louis Fieser. When Alex visited Syntex in 1950, he was charmed by the makeshift laboratories in which open windows and doors substituted for fume hoods, and explosive reactions were performed in an adjacent patio. And the work at fever pitch had just culminated in winning the race, against world-famous teams from Harvard and Merck, to synthesize cortisone from diosgenin. As recounted in *Harper's* magazine: "Big minds rather than

Alex in Mexico, with a 20-kilogram dioscorea root unearthed in the jungle

big research budgets, lead to big discoveries. . . . Last, but not least, . . . the leader in the race was a chemical manufacturer in presumably backward Mexico."

Syntex's main business, when Alex came on board in 1951, was to furnish the steroid building blocks that pharmaceutical companies needed for making their hormone products. The crucial element became the supply of Mexican yams, the raw material from which the diosgenin was extracted. Alex wondered whether there would be enough raw material in Mexico for the next ten years and whether a botanical studies program could identify other sources of diosgenin in Central America and South America. How could puny Syntex compete in a battle with multinational giants, the likes of Upjohn and Merck?

Alex began by organizing a laboratory close to the Veracruz jungles, from which he could obtain a supply of fresh roots to work with. The terrain was difficult, the part-time workers from the sugarcane plantations were far from reliable, and processing of the roots was primitive and haphazard. The yam roots were chopped up with machetes and left in the fields for weeks to soak in the rain and dry under the sun. The diosgenin yield was pitifully small and erratic. Yet Somlo had warned Alex: "It's impossible to organize any work in the Mexican jungle. You cannot get capable people to work there. You should be satisfied that we have this contractor to do the root collecting for us."

But Alex was persistent, and Somlo finally gave him full authority over the project. Alex found a wonderful Spaniard, Don Emilio Fortanet, who helped him set up and run second-hand equipment that had been intended for chopping and drying potatoes. The roots processed with this equipment were then finely ground, extracted, and bathed in a solution containing the known precursors for the biosynthesis of steroids. With these and other refinements, Alex improved the yield of diosgenin manyfold and made the procedure dependable at a far lower cost. The success of this operation gave Alex a unique sense of fulfillment. Contrary to the advice of older, experienced mentors, Alex had recruited local laborers and imbued them with a team spirit to advance the technology by their efforts. By applying the scientific insights acquired in his graduate research at Rochester, he had made a real impact on a practical problem. A dual career as scientist and entrepreneur would come to suit his tastes and talents.

Turning Syntex into a Pharmaceutical Company

Syntex was first in the race for steroid synthesis on an industrial scale. They supplied progesterone to Upjohn by the ton for use in the microbial conversion to cortisone, which later became a billion-dollar business. Even so, Syntex was not doing well. Beset by falling prices and by legal problems and other disputes, Somlo decided, in 1956, to sell the company to Charles Allen, a New York financier. Against his lawyer's advice, Allen bought Syntex because of his faith in its scientists and the belief they would soon develop an oral contraceptive to curb the world's exploding population. He created Syntex Corporation as a Panamanian company to gain tax advantages, and he placed its management in the hands of Rosenkranz and Zaffaroni. They, in turn, persuaded Djerassi, who had left in 1951 for a professorial post at Wayne State University in Detroit, to return as Director of Syntex Research. (Allen's astute acquisition of Syntex, combined with the phenomenal capabilities of these three technical giants, resulted in an increase in its market value from less than $5 million in 1956 to more than $1 billion a decade later.)

The Rosenkranz–Djerassi team of bright and energetic chemists synthesized a variety of hydrocortisone-like compounds, which were tested in animal models for hormonal activity by Ralph Dorfman at the Worcester Foundation. But the labor and time needed to make each of these compounds limited their number and range. As an alternative biochemical approach, Alex used adrenal extracts to perform the difficult introduction of the signature unit of steroids, the hydroxyl group in position 11-β. Now a very much larger number of synthetic compounds could be assayed; those with a fluorine atom replacing hydrogen in position 6 were found to be especially active against inflammation—a breakthrough in therapy that became a marketing success as Synalar.

Alex had realized early that, as a bulk supplier of steroids, the future of Syntex was limited, compared with a company making and marketing medicinal products. The road to becoming a fully integrated pharmaceutical company was mined with traps, but there really was no choice: it had to be. By 1957, they all agreed that, with the increasing competition in the limited bulk-steroid field, Syntex had to take the pharmaceutical route. Unfortunately, Syntex's resources were too limited. Alex took increasing respon-

Alex with Miguel Alemán Valdés, then president of Mexico, setting up the first birth-control clinic in Mexico

Alex with Carl Djerassi in 1954

Alex in Mexico, with Dr. George Rosenkranz (left) and Syntex attorney Lic. Licio Lagos

sibility for negotiating deals with Eli Lilly, Ortho Pharmaceuticals, and other companies to fund Syntex research in exchange for marketing rights to future products. He also headed the pharmacologic and clinical testing, which would lead to FDA approval for Synalar and norethindrone, an oral contraceptive.

In the development of Synalar, Alex made three key decisions. One was to replace the adrenal-extract procedure, which entailed arduous collections of 100 kilograms of glands at a time from slaughterhouses with only feeble yields of the product. To this end, Alex and his team searched for and discovered a Mexican soil bacterium that could perform the same chemical operation as the adrenal extracts but far more easily and cheaply. The second decision was to conduct a clinical testing program that exceeded FDA standards of safety and efficacy for a broad range of skin diseases. The third critical decision stemmed from an observation made by a clinical consultant that the drug was more effective against psoriasis lesions when covered with Saran Wrap. This concept, a forerunner of the sophisticated ALZA therapeutic systems, was exploited in Synalar, and it enabled Synalar to capture the market from a rival Eli Lilly product promoted by a sales force that was well-known and respected in the industry and ten times the size of Syntex's sales force.

By 1960, Alex, now President and Chief Executive Officer of Syntex Laboratories, saw the need for a location in the United States for clinical testing, corporate affairs, and marketing. Strict visa regulations in Mexico, which precluded permanent employment of noncitizens, restricted the expansion of research and development. New York and its New Jersey environs were the center of the U.S. pharmaceutical industry, but Alex was unhappy with the prospects of metropolitan commuting and intemperate weather. Instead, with Djerassi now at Stanford, the Stanford Industrial Park seemed preferable—with its equable climate, tranquil ambience, proximity to the university, and easy access both to Mexico and to New York. Alex and Rosenkranz managed to get the university to lease them 100 acres of lovely rolling hills for the future site of a Syntex campus, the first chemical–pharmaceutical company in the park.

With responsibility for Syntex research as well as for development, Alex conceived an organization that would offer an academic appeal to top-flight scientists. There would be separate institutes for each of the disciplines: institutes of steroid chemistry, steroid biology, pharmacology, and pharmaceutical sciences, as well as an utterly novel Institute of Molecular Biology.

In 1961, the Institute of Molecular Biology was the first Syntex research unit established in Palo Alto. With the guidance of Djerassi and Joshua Lederberg, a Syntex advisor also from Stanford, the institute was directed by John Moffatt in organic chemistry, by Boris Rotman in microbial genetics, and by Bill Razzell in enzymology, and it was staffed by a total of fifteen scientists. John Moffatt had done his graduate work on the synthesis of phosphate compounds under H. Gobind Khorana at the University of British Columbia, the first of the very few graduate students ever trained by Khorana in his illustrious direction of hundreds of postdoctoral fellows. John, separated from exciting research in his first job with Calbiochem (then in Los Angeles and now in La Jolla), had been miserable for nearly a year, but he was reborn with the opportunities at Syntex.

Among the projects explored by the molecular biology group were the effects of nucleoside analogs on nucleic acid biosynthesis and on cell growth. The dideoxynucleotides (much the same as the nucleotide building blocks of DNA, but lacking the chemical group that would allow them to be linked into a chain) that emerged from their work were invaluable in my studies of the mechanisms of DNA polymerase action, which, in turn, inspired Fred Sanger to use them in his celebrated procedure for sequencing DNA. These nucleotide analogs also became the basis for the drugs (such as AZT) designed by Burroughs Wellcome and Syntex that are now used in the treatment of AIDS and certain other viral diseases.

Around 1970, after Alex had left to found ALZA, the management of Syntex abolished the Institute of Molecular Biology. Moffatt was put into the large Division of Organic Chemistry, and John Josse, an outstanding enzymologist who had left a professorial post to replace Razzell, became disillusioned with the new research direction and entered medical practice. Only a few years later, the discoveries of recombinant DNA, and the sequencing and cloning of DNA, catapulted molecular biology into the forefront of the biotechnology era. Had the early vision and encouragement remained to preserve that modest unit, these new technologies would surely have been exploited by Moffatt and his colleagues and would have put Syntex years ahead of Genentech and others in the genetic-engineering race. Instead, the embers of molecular biology were smothered by the traditional pharmaceutical practice of synthesizing thousands of compounds for routine screening in the hope of finding a drug that is active in some disease model.

Founding of ALZA

By 1968, Syntex was rich and successful, the only company in the post–World War II era to break into the profitable pharmaceutical business. It made its fortune by the classic route of organic synthetic research that yielded Synalar, a steroid skin cream used against a variety of allergic, psoriatic, and other dermatologic disorders, and "the pill," the first effective oral contraceptive made of norethindrone, a progestational compound. Each was a blockbuster product with a major share of its market. Synalar became the leading topical anti-inflammatory product, and norethindrone became the leading compound in oral contraceptive markets (with

several pharmaceutical companies using it in birth-control products sold under different brand names).

Like the management of many other big and successful companies, the directors of Syntex had grown comfortable with what
they were doing and very cautious about pursuing new, unproven
ideas. Profits were accumulating, but how to invest them? Innovative and ambitious programs launched within the company would
lose out every time in competition with ongoing and entrenched
profitable activities. The company recognized that, for a novel direction to have a chance, it would have to be launched outside the
company; but there were different views on how this should be
done.

The Syntex board of directors, intent on diversifying and expanding in the health-care field, favored the acquisition of established companies in diverse lines of business: dental and veterinary
products, cosmetics, and instrumentation. Several such investments were made. The acquired management was retained, but
Syntex assumed control. After some years, when each of these ventures proved unprofitable, they were divested by Syntex management to the applause of Wall Street analysts. By contrast, Alex favored spin-offs, nucleated by ideas and staff from Syntex, initially
financed and partially owned by Syntex. The new venture—physically, financially, and culturally independent—could then embark
with the resources to make it on its own.

Alex's strategy was followed in 1966, when he and Djerassi arranged for Syntex and Varian Associates to join in creating Syva
Corporation to exploit the free-radical (electron spin resonance)
technology pioneered by Harden McConnell at Stanford. With advice from Avram Goldstein, these techniques were applied to meet
the urgent need for rapid drug testing of American forces in Viet-
nam. The technology has since been expanded into a highly profitable medical-diagnostics business. Yet, when Alex argued for a
spin-off to exploit the steroid insect molting hormone ecdysone
for pest control, Syntex management was resistant. Eventually,
however, they let Djerassi start Zoecon, a company with that very
objective.

Persuading the Syntex board to start a company dedicated to
long-range programs to innovate drug-delivery systems seemed
hopeless. While George Rosenkranz and Charles Allen seemed
supportive to Alex in person, the board they controlled would

never take on the risk of such a bold new venture. Alex resolved to leave Syntex in 1968 and to start a new company that would integrate the discovery of novel drug-delivery devices in the laboratory with their ultimate acceptance by doctors and patients worldwide. To mark his total commitment to this grand scheme, he pledged the company his name, *AL*ejandro *ZA*ffaroni, sold his Syntex holdings, and invested half his fortune for the initial financing of ALZA.

For many years, Alex had believed traditional pills and injections to be archaic. From his early and continuing work with adrenocortical steroids, he was aware of the limitations of hydrocortisone therapy. The serious side effects of the original, several-times-daily dosage had not been lessened by sophisticated manipulations of chemical structures (although a change to a single daily dose was later developed). In the normal body, this and other hormones require refined controls for their production and release into blood and tissues. How could one expect them to function optimally when swallowed or injected at arbitrary intervals? Within moments of administration, conventional pills promptly disintegrate. The resulting quick absorption of the drug often produces a sharp peak in its blood level, contributing to side effects, followed by a rapid fall to levels ineffective for therapy.

Alex's plan was to develop devices that would make the arsenal of discovered and proven hormones and drugs, including those still under patent protection, far more effective and better tolerated. The ALZA therapeutic system would incorporate three features: First, the temporal patterns of a particular drug's interactions with its target would be determined and used to design an optimal dosage scheme. Second, bioengineered drug-delivery systems would be developed to carry out the program. Third, Alex envisioned a means of delivering a drug exactly where and when it was needed and of maintaining it there at an efficacious level, potentially avoiding adverse side effects.

Eventually, patents were issued to Alex for delivery systems that would release drugs at a constant and controlled rate for a prescribed length of time—a week, say, or a year. These were the first of many patents for ALZA. In contrast, conventional pills or injections produce a brief peak of an excessive level of drug, often associated with side effects, followed soon, upon the body's elimination of the drug, by a long trough of drug levels that may be subtherapeutic. Compared to slow-release capsules and other available crude devices, the ALZA systems would be (to use the metaphor of that era) the "Polaroid" of pharmacology.

To institute the drug-delivery program, systems would have to be designed, developed, and tested; then they would have to clear the hurdles of extensive clinical trials, regulatory approvals, and the minefields of marketing. It could take many years, cost one hundred million dollars or more, and still fail at any stage. At first, Alex did everything himself—he sought financing, recruited personnel, planned and built facilities—all with taste and dispatch. A key choice, as Chief Financial Officer, was Martin Gerstel, who, at age 28, had just completed his MBA with highest honors at Stanford. Gerstel was so convinced and charmed by Alex and his vision that he accepted the job on the spot, without even knowing his title or salary.

Two conditions were initially selected for which a system could markedly improve the delivery of a known drug to an accessible organ. One was glaucoma, to be treated by an ocular insert to deliver the pressure-relieving drug pilocarpine to the eye. The other was prevention of pregnancy by the release of progesterone from a small, T-shaped device placed in the uterus. The essential principle of each delivery system was that drug molecules, contained in a reservoir made of a polymer membrane, diffused through the membrane and were released at a controlled rate over an extended period of time. The Ocusert system (Figure 2) would deliver pilocarpine for one week; the contraceptive product, Progestasert (Figure 3), would release progesterone for one year.

In addition to the many technical problems to be solved, some basic pharmacologic questions, despite decades of clinical experience, needed further study—the optimal level of a particular drug, its persistence in body compartments, and side effects. The delivery system required that controlled chemical syntheses be developed to obtain polymers with the proper permeability and release characteristics. In designing the system, it was essential that it be easily inserted and removed, that it be comfortably retained without irritation, that the drug delivery be exact and fail-safe, and that the packaged device be sterile and stable on a shelf for years.

Technical issues became intertwined with financial and regulatory concerns. As a part of the separation agreement between Syntex and Zaffaroni, Syntex, in 1968, had obtained approximately 25 percent of the original issue of ALZA common shares. However,

Alex with Martin Gerstel, Co-chairman and CEO of ALZA

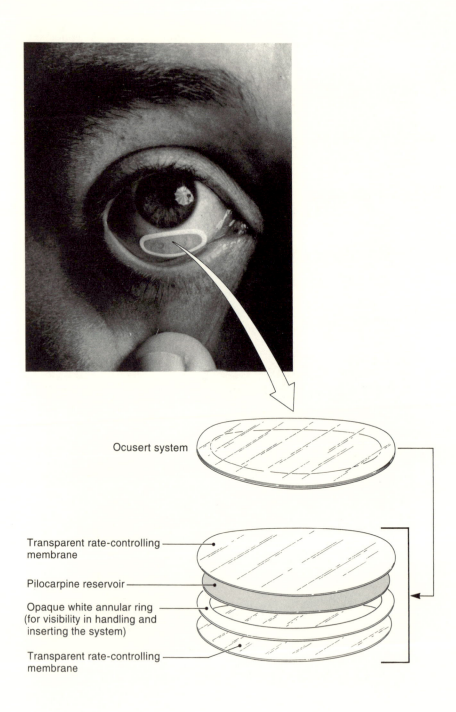

Ocusert system

Transparent rate-controlling membrane

Pilocarpine reservoir

Opaque white annular ring (for visibility in handling and inserting the system)

Transparent rate-controlling membrane

FIGURE 2. The Ocusert system in the eye (*facing page*) and the components of the Ocusert system (*bottom*). Ocusert systems deliver pilocarpine at the rate of 20 or 40 micrograms per hour for 7 days. The dimensions of these two release-rated systems are 13.4 mm × 5.7 mm × 0.3 mm (for the 20-µg system) and 13 mm × 5.5 mm × 0.5 mm (for the 40-µg system). The Ocusert system was the first product that the FDA ever approved with label specifications for rate and duration of drug delivery.

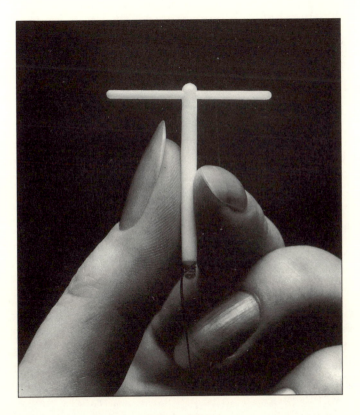

FIGURE 3. The Progestasert system delivers about 65 micrograms of progesterone to the uterine lining every day for one year. The length of the progesterone-containing stem is 36 mm; its outer diameter is 2.8 mm. The crossarm is 32 mm long, and its diameter is 1.6 mm.

it soon became apparent that Syntex's ownership position could create problems for ALZA. In particular, the Progestasert contraceptive system, if successfully developed, would compete directly with Syntex's birth-control pills. With this in mind, ALZA and Syntex agreed that Syntex would distribute all of its ALZA shares to Syntex's shareholders as a dividend.

But the United States Securities and Exchange Commission (SEC) and the various state securities agencies were not then in the mood to be agreeable. Although Syntex's shareholders were not being asked to make any dollar investment to receive their ALZA shares, the SEC would not allow the public distribution of the ALZA shares. The reason for their concern was that it was generally considered unacceptable for a company to have a public market for its securities before it had earnings and a reasonable expectation to continue operations for the foreseeable future. ALZA, of course, had no revenues, its expenses were growing, and its funds (a substantial portion of which had been provided by Zaffaroni himself) were very limited. Inasmuch as no additional funds were being raised from the public, regulatory approval for distribution of the Syntex-owned ALZA shares was finally received.

In late 1969, ALZA became a public company with thousands of shareholders and a market value approaching $100 million. At that time, the company consisted of a few employees, limited assets, and Alex's vision for revolutionizing drug delivery, but no actual product—and, of course, no means to judge if or when it would ever have sales or profits. The public trading of ALZA securities, at such an early stage in its corporate life, was a precedent that would be exploited in the following years for the establishment, growth, and survival of many biotech ventures.

From the time of the Syntex distribution of the ALZA shares in late 1969 through early 1976, ALZA further financed the company through a number of private and public offerings. As work proceeded rapidly on many fronts, the excitement and fever often spread to Wall Street. Analysts knew little about polymers, but they liked the sound of them and believed that their use in novel medical devices would produce large profits. Stock traders believed that they could profit from the sizzle of the story.

The immediate worry at ALZA during the first half of the 1970s was imminent competition from the pharmaceutical giants. The fear was that progress at ALZA would alert them to enter the race and that they would quickly overtake fledgling ALZA. These

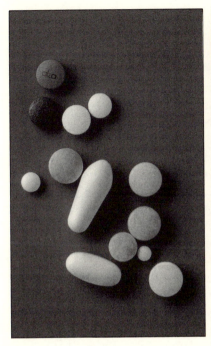

FIGURE 4. OROS systems, like conventional pills, come in various sizes and shapes (*left*), but they deliver drugs only in liquid formulations. Here (*right*), a dye is flowing from an OROS system into water to show how a drug solution or suspension would flow out of the system at a controlled rate. The OROS Elementary Osmotic Pump (the simplest OROS system) delivers soluble drugs, by means of an osmotic process, for up to 24 hours. Each system consists of a semipermeable membrane enclosing a drug-containing core. In an aqueous environment, such as the gastrointestinal tract, water is imbibed through the membrane at a controlled rate, causing the drug to go into solution. The solution flows out of an orifice in the membrane at the same controlled rate at which water is flowing into the core.

concerns proved unfounded as it became obvious that the giants couldn't be force-fed new technology, even after a decade of trying to persuade them.

The price of ALZA's common shares zigzagged up and down. Financial analysts attempted to project what this new technology would mean in the pharmaceutical world. By 1974, investors began to lose hope, as the company continued to show increasing losses and there seemed to be good reason to be concerned about its viability. Then, however, Ocusert systems were cleared for marketing in 1974, and Progestasert systems received preliminary approval in 1975, reviving the company's prospects. ALZA's stock rose from a low of $7 in 1974 to a high of $30 in 1975.

One day, I remarked to Gerstel that, when the Ocusert and Progestasert systems finally went on sale, the stock price would go through the roof. He wisely cautioned that the reverse might happen—often, perversely, the financial community reacts negatively to success, in that its expectations for performance had been grossly exaggerated.

The ALZA divisions of polymer chemistry, physical chemistry, pharmacology, engineering, and manufacturing all met their goals and blended nicely to produce devices that were ready for clinical trials. The FDA was patiently and effectively educated in the novel pharmacokinetics (that is, drug-level maintenance) and the protocols to evaluate performance of the delivery systems. Ophthalmologists were identified to conduct the Ocusert trials on glaucoma patients, as were gynecologists for the Progestasert trials for contraception.

The Ocusert gave glaucoma patients great relief from the burden of taking pilocarpine eye drops four times a day. Worse than the nuisance of getting the drops into the eye was being immobilized for an hour or more until the pupils, constricted by the drug, recovered enough that proper vision was restored. No wonder that compliance with the eyedrop regimen had generally been erratic. Yet there had been no practical alternative: failure to lower the intraocular pressure eventually produced blindness. In contrast to eyedrops, the Ocusert, once in place under the eyelid, delivered the pilocarpine unobtrusively for exactly a week, and, for many patients, it kept the intraocular pressure under control far better than did the eyedrops. However, annual sales of the Ocusert systems never approached the original projections.

At the dedication of the ALZA Institute for Pharmaceutical Sciences in Lawrence, Kansas, in 1969: (from Left) John Shell, the author, Hans Selye, Alex Zaffaroni, and Takeru Higuchi

What went wrong? Clearly, success in science had not lead to success in the market. Putting a centimeter-long object in your eye can be intimidating. The ease and enthusiasm with which volunteers on the ALZA staff could insert and retain the Ocusert was not common among older people with glaucoma. These patients needed training and sustained encouragement from their ophthalmologists, few of whom would take the time and devote the necessary effort. To make matters worse, Merck released Timoptic, a new drug, in the form of eyedrops, that was as effective as pilocarpine but that required administration only twice a day. The Merck product now has annual sales of more than $150 million, while the Ocusert system still has only a very small part of the market.

The Progestasert system is a tiny unit. It contains, in its matchstick-size stem, enough progesterone to last a year. When it is placed in the uterus, infinitesimal amounts of hormone (about 65 micrograms per day) exude from the device, but these amounts are sufficient to achieve effective contraception. Insertion of the device by the gynecologist takes only a few minutes, and the presence of the device in the uterus usually does not interfere with normal menses. Few women who use the device complain of bleeding or discomfort; often, in fact, women find relief from menstrual cramps with the Progestasert system.

Unlike the "pill," the device produces no systemic effects, because no significant amount of hormone goes beyond the uterus. Patient compliance, always a problem with self-medication, is avoided, except for the annual replacement. To this day, units are sold to many people who properly regard the Progestasert as the ideal contraceptive. Why, then, did it fail to obtain a significant market share?

From the Brink of Bankruptcy

The Progestasert system was introduced at the very time that the Dalkon Shield, a crab-shaped plastic intrauterine device, had become the notorious subject of countless complaints and huge class-action lawsuits. Infections and other adverse reactions caused by the Dalkon Shield raised fears, which were widely publicized, about the safety of this and other intrauterine devices on the market. Progestasert, though designed as a drug-delivery unit, had the burden of being classified, by the Food and Drug Administration (FDA), as an intrauterine device and, thus, shared a reputation that had been made by entirely different types of devices.

An inherent limitation of the Progestasert system cropped up and attracted exaggerated attention. Because the device blocks neither ovulation nor fertilization, it cannot prevent an ectopic (tubal) pregnancy, which occurs in about 2 percent of all pregnancies. The relative risk of tubal pregnancy increases with age and with pelvic inflammatory disease, such as a venereal infection. Hence, women who are positive for either of these risk factors are unsuited for the Progestasert system, as the package insert emphatically warns.

In retrospect, Alex and his marketing chief, Carroll Walter, might have foreseen the problems with the Ocusert system, but the Progestasert mishaps could be blamed only on bad luck and bad timing. Up until this time, the thrust and focus of ALZA was on research and technology, with marketing being only an appendage. Alex realized this shortcoming, and, around 1973, he was close to concluding a merger with Marion Laboratories in Kansas City, Missouri, a company superbly successful in marketing licensed drugs but lacking the pipeline to develop new products, which ALZA could have amply filled. Unfortunately, the deal collapsed, largely because Wall Street analysts did not think well of it and discouraged stockholder approval.

The marketing problems that initially threatened the survival of ALZA reminded me of a parallel interplay in the microbial world. Mutants of *Escherichia coli* were discovered that were defective in cell division, such that they sometimes pinched off minicells entirely devoid of DNA. Despite lacking the "intelligence" and direction of DNA that is necessary for producing future generations, the minicells were still very much alive. They were able to respire and to engage in a lively and profitable energy exchange with nutrients in their environment. For someone with a vested interest in DNA as the molecule of heredity, I was reassured when I learned that the DNA-less cells had no reproductive future. By the same token, we know that DNA, by itself, without the cellular machinery to sustain and implement its vast store of information, can provide neither a present nor a future.

These observations made me aware, in a broader philosophic sense, of the relative values of ideas and information, on the one hand, and their implementation on the other. Drugs and devices, like ideas and discoveries, need not only to be tested and used successfully but also to be marketed and accepted on a wide scale, in order for their true value to be established. Simply *having* a good idea or a good product will not make it sell. Unfortunately, the reverse is also often true. Marketing can be successful, for long periods, even though it may lack a sound foundation. This power of marketing can be found in all precincts of human behavior, including the practice of medicine and the pharmaceutical industry.

The technological triumphs of the Ocusert and the Progestasert systems were soon overshadowed by the development of a new OROS delivery technology: pills were created that, through the use of osmotic energy, essentially functioned like tiny pumps (Figure 4). Drugs incorporated in the core of an osmotic pill could be delivered, at preselected rates, through a laser-drilled hole as the pill transited through the stomach and intestines. Not only could this dosage form be used to release a drug at a controlled rate over an extended time period, it could also be used with many different chemical entities. One technology could lead to numerous new pharmaceutical products (Figure 5).

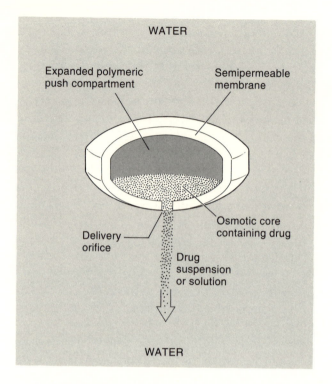

FIGURE 5. The OROS Push-Pull Osmotic Pump system delivers highly soluble or insoluble drugs for up to 24 hours. Each system consists of a semipermeable membrane enclosing a two-layer core, one layer containing one or more drugs and excipients and the other containing osmotic agents. In an aqueous environment, such as the gastrointestinal tract, water is imbibed through the membrane at a controlled rate, causing the osmotic layer to swell and the drug to go into solution or suspension. As the expanding osmotic layer pushes against the drug layer, the drug formulation flows out of the drug layer, via an orifice in the membrane, at the same controlled rate at which water is flowing into the core. Different drugs are deliverable at different rates from the same system.

A second technology depended upon the development of very sophisticated systems—which superficially resembled small adhesive bandages—to deliver drugs through unbroken skin (transdermally) into the blood stream (Figure 6). This mode of drug delivery was sometimes referred to as "an infusion without a needle." These systems, popularly called "skin patches," incorporated a drug reservoir, a rate-controlling membrane, and adhesives, which together allowed for the controlled transport of a drug to and

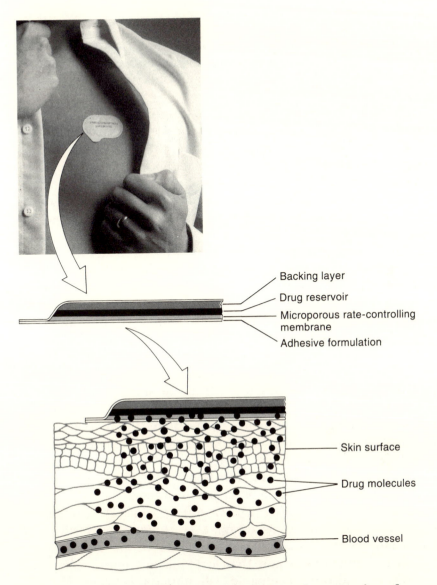

Backing layer
Drug reservoir
Microporous rate-controlling membrane
Adhesive formulation

Skin surface
Drug molecules
Blood vessel

FIGURE 6. Many ALZA transdermal therapeutic systems have four layers: an impermeable backing to prevent drug evaporation; a drug reservoir; a membrane to regulate the release rate of the drug; and an adhesive to attach the system to the skin. When a system is placed on the skin, the drug begins flowing from the reservoir, through the adhesive, and into and through the skin into the capillaries under the adhesion site and, thence, into the systemic circulation. A loading dose can be incorporated into the adhesive to bring the blood concentration of the drug up to therapeutic levels quickly, before the slower, membrane-controlled release from the reservoir has had much effect.

through skin for as long as one week—for example, to control high blood pressure. To expand the opportunities for transdermal delivery of drugs, chemical penetration enhancers were developed to increase the variety of drugs that could be administered by this route.

Though discoveries abounded, one intractable problem remained: how to persuade a pharmaceutical company to utilize one of ALZA's technologies to improve the performance of a profitable drug or to extend its patent life. The "not-invented-here" syndrome is extraordinarily powerful and prevalent in the pharmaceutical industry. The only way, it seemed, was for ALZA itself to select a drug and to demonstrate the effectiveness of improved delivery with one of its devices.

But clinical testing is expensive, and ALZA lacked the resources. With mounting annual losses from the unsuccessful commercialization of Ocusert and Progestasert systems, and with its growing research-and-development (R&D) expenses far exceeding revenues, ALZA was at the brink of bankruptcy. On several occasions, the company was saved by an eleventh-hour bank loan or by a financial maneuver cleverly devised by Alex and Martin. Once, when things were especially grim, Alex convened a retreat with his top staff to decide what to jettison from the sinking ship. Even in this extreme condition, they adhered to the principle of building long-term value. The company's early attempts at marketing and manufacturing might have to go, but research would be preserved to the very end. Nevertheless, contracting the company's horizons, so that it was no longer an integrated pharmaceutical company, but an R&D organization, was a bitter pill to swallow.

I recall a social evening in 1977, during which Alex remained outwardly unperturbed, as always, even though foreclosure of a bank loan was only days away. He and his top people had made extensive presentations to more than forty pharmaceutical, chemical, petroleum, and other companies, all, apparently, to no avail. Finally, two companies showed serious interest. Atlantic Richfield, the petroleum company, was considering diversifying into the medical field. Ciba-Geigy, the pharmaceutical company in Basel, Switzerland, having decided to embark on a drug-delivery research program, realized that ALZA was years ahead of them, with an exceptional group of scientists and a broad base of proprietary technology and patents.

The deal Ciba-Geigy offered ALZA in late 1977 was painful, but there was no choice. ALZA's earlier fears of competition were

Martin Gerstel

in essence realized in this transaction; the company and its future were being devoured by a pharmaceutical giant. For $30 million, Ciba-Geigy acquired preferred ALZA shares representing 80 percent of the voting rights of the company (and convertible into 50 percent of the total equity), giving Ciba-Geigy total control of the company. Funds would be provided to transform OROS and transdermal technologies from laboratory status to the manufacturing level. In addition, under an R&D agreement, Ciba-Geigy would fund research at ALZA at least to the tune of $15 million, and, in return, would receive rights to all of the products developed and exclusive right to all of ALZA's existing and future technology.

The value of an ALZA share, a sad reflection of its market worth, had dropped to about $2 to $3 per share from the $40 per share of only a few years earlier. Technological success had not been matched in the market, nor had a safety net been built of products marketed by others. Understandably, some investors who lost were disgruntled and angry with Alex. But Alex had never touted the stock; he had kept his own shares throughout, and he had done his utmost to enhance their value.

Despite serious problems that developed later during ALZA's brief marriage to Ciba-Geigy, it accomplished its original purpose: to save ALZA from extinction. Given the lack of alternatives, the move was inevitable, and both sides ultimately benefited. One positive aspect was that ALZA did not feel threatened by Ciba-Geigy in a technical sense because that company's technical people were not interested in what ALZA was doing. In fact, no technology transfer occurred.

Nevertheless, ALZA found itself closely controlled just when opportunities for its expansion were suddenly developing. A few forward-looking pharmaceutical companies had grown interested in drug delivery and approached ALZA about possible collaborations. ALZA pleaded with Ciba-Geigy for some freedom to reach out to those companies to pursue the increasing opportunities, but ALZA's pleas were rebuffed. The controlling partner maintained the contractual restraints and was never comfortable with ALZA's chafing over the monogamous marriage or with the research and managerial style of its California acquisition. Nor were the Ciba-Geigy scientists receptive to the enormous potential of the emerging ALZA technologies.

In late 1981, after four years of a proper but awkward relationship, Ciba-Geigy agreed to relinquish its control of ALZA. The ALZA investment ultimately proved highly profitable for Ciba-Geigy. The acquisition enabled Ciba-Geigy to write off the large pre-existing and ongoing ALZA losses against their American profits and to effect a tax savings that nearly matched their investment. Their equity holdings in ALZA later increased in value to hundreds of millions of dollars, and their accelerated entry into the drug-delivery market provided them with worldwide cumulative sales of ALZA-developed products in excess of $3 billion by the early 1990s. Still, had they been more understanding and more given to openness and patience, Ciba-Geigy could have profited much, much more.

At last, ALZA was on its own again. By now, many pharmaceutical companies had begun to recognize the value of drug-delivery systems. A favorable analyst's report in 1982 drove ALZA stock from $2 per share to $9 per share. Imitators with less sophisticated devices began to establish highly profitable businesses. Key Pharmaceuticals in Kenilworth, New Jersey, with an energetic sales force promoting a relatively crude transdermal nitroglycerin patch and a sustained-release oral product for asthma, was acquired in 1986 by Schering-Plough with stock worth $800 million.

The ALZA therapeutic systems were now in great demand. By the end of 1993, ALZA had agreements for the development of more than sixty products and royalty income from products already commercialized in excess of $110 million per year. Sales of ALZA-developed products exceeded $2 billion per year, with one product, Procardia XL (nifedipine), marketed by Pfizer, surpassing $1 billion in annual sales. ALZA's total market value, which had plummeted to less than $20 million at the time of the separation from Ciba-Geigy in 1982, climbed, after two stock splits, to more than $3 billion by early 1992.

Defeat at Dynapol

ALZA focused on the use of polymeric membranes, in the form of patches or enveloped pills, as devices for the controlled release of drugs. With ingenious adaptations for a variety of drugs, new ways had been uncovered to lower blood pressure, to decrease angina attacks, to reduce chronic severe pain, to prevent motion sickness, to provide hormone replacement, and to treat periodontal disease.

More ideas kept surfacing to meet still other health problems. Among them were ways to meet the threats of disease attributed, in the early 1970s, to the use of food additives. The food industry was faced with the need to replace the coloring agents, sweeteners, and antioxidants that had been banned or were (fairly or not) suspected of causing health problems. The FDA had just prohibited the use of cyclamate because a few test animals given astronomical doses for several years had developed cancer. A law passed by Congress, known as the Delaney amendment, mandated the prohibition of any drug or agent, no matter what the dose, that had been shown to cause cancer in an animal.

Alex was intrigued by a published report that insulin linked to a polymer could exert its normal hormonal action by binding to a cell-surface receptor, while remaining outside the cell. Could a polymer be used in a similar way to render a food additive unabsorbable? A polymeric leash attached to a dye, a sweetener, or an antioxidant molecule might prevent its absorption from the intestinal tract without altering the properties of the additive. Having colored or flavored a food or having removed an oxidant, the additive would later be completely eliminated in the stool.

Despite the apparent simplicity and feasibility of the idea, a program to implement it would have to compete for funds and space with important ongoing projects at ALZA. Even more essential to Alex in this, as in a number of like situations, was that a new initiative be given undiluted attention in a fresh environment. Starting with two ALZA staff members, additional scientists and supporting personnel would be recruited and given free rein to create their own patterns and procedures, and they would be asked to make a commitment to only one goal—the success of the new venture.

Dynapol was started in 1970 in the Stanford Industrial Park near ALZA in a building remodeled into attractive laboratories and offices. Investment by private individuals raised $18.5 million. Steve Goldby, selected as the CEO, along with Ned Weinshenker, the research director, assembled a staff of more than a hundred, including able young chemists, pharmacologists, and business-people. They soon succeeded in the chemical synthesis of polymeric appendages to food additives that made them virtually unabsorbable during their passage through the gut. A minor exception was the uptake of a small amount of dye by the intestinal lymphatic system after ingestion of amounts in huge excess over what an average human might consume.

Like ALZA, Dynapol, despite meeting its technologic objectives, foundered for lack of the financial support needed to weather regulatory and marketing uncertainties. Start-up funds had been provided by private investors recruited by Alex. But the major infusion of capital, nearly $20 million, was from the DeKalb Company, a large midwestern agricultural firm that was also in the oil business. With this investment, DeKalb assumed control of Dynapol and held its fate.

Saccharin, which was being substituted for the outlawed cyclamate, had also been shown to be carcinogenic. Yet the anticipated ban on its use never materialized. Now the higher cost of producing polymer-leash sweetners would make them less competitive with saccharin. Faced with financial reverses in its own business, DeKalb lost confidence and bailed out. Dynapol was finished. Ironically, the FDA did approve the Dynapol antioxidant, but DeKalb let it languish. What started out as a good idea, and had been well executed in the laboratory, could not get past the rocky shoals of regulation and marketing in the real product world.

Lure of Biotechnology: DNAX and Affymax

By the late 1980s, ALZA was safe and secure. A handsome royalty stream flowed from many pharmaceutical companies in return for ALZA-developed drug-delivery products already on the market, with many more products and systems in the pipeline. With Martin Gerstel, Alex's closest associate, installed as the chief executive officer, ALZA operations were in able hands.

DNAX had also matured by this time, having become an acknowledged leader in molecular immunology. Cytokines, many of them discovered at DNAX, were active in the immune and hematopoietic systems and had been chosen for development as drugs by Schering. The success and maturity of these ventures left Alex with much less to do, and, to his friends, he seemed visibly dispirited. Diversions, including extended visits to Uruguay, could not replace the excitement of a new venture, and this he finally found by creating Affymax to fulfill still a different mission.

At DNAX, the mission was to discover the body's own macromolecules that govern the growth and development of cells and tissues and to produce them in quantities sufficient to correct diseases of the immune and blood-forming systems. At ALZA, the goal was to enhance the value of known drugs, which, as small molecules, could be made far more effective by rational delivery systems. But more of these drugs were needed, and the process of their discovery had become increasingly difficult and prohibitively expensive. Between 1970 and 1990, the number of new drugs approved by the FDA remained constant at about twenty per year, but the average cost of developing a drug had risen from $500 thousand to more than $300 million. What was sought in the pharmaceutical industry, and what Affymax was designed to supply, was a rapid, less expensive way of identifying effective drugs among the untold numbers of molecules already in hand or newly created.

Effective treatments for the myriad diseases may already exist among the hundreds of thousands of compounds synthesized by companies over many decades and among the vast number of natural small molecules in soil and plants. To these could be added the countless variety of short chains assembled from amino acids (that is, peptides) or from nucleic acid building blocks (that is, oligonucleotides). With new technologies derived from recombinant-DNA research and with highly automated, miniaturized techniques to detect the reactivity of small molecules with cell-surface receptors, antibodies, and other natural macromolecules, the time and cost of discovering attractive drug candidates might be reduced by orders of magnitude.

At Affymax (named for *affinity* *ma*trices), Alex was once again launched into the feverish activity of raising money, organizing an elite scientific advisory board, recruiting scientists and staff, and building laboratories. In addition to the technological and business experiences of his past, he reached out for the electronic innovations of data recognition, data storage, and data retrieval available from neighbors in Silicon Valley. While other biotech ventures have similar objectives and access to similar technologies, only Affymax has Zaffaroni. Given his track record and verve, he has generated several joint ventures with pharmaceutical companies and convinced investors to invest $184 million, of which $92 million came in 1991 from an initial public offering. With a proven formula of innovation, technology, quality, and persistence, Alex obtained the resources to try to make it happen again. In January 1995, five years after its founding, Affymax was acquired by Glaxo, Inc., for $533 million, an acknowledgment by the drug-company giant that the future discovery of drugs will be accelerated by unique combinations of electronics and biotechnology. Affymax's innovative use of miniaturization methods borrowed from semiconductor industrial neighbors in Silicon Valley make it possible to synthesize vast numbers of different compounds on tiny silicon chips and microscopic beads and to screen them for potential medical uses.

Attitudes and Philosophies

"You have to be a gambler to start a new company." Alex, having started many companies, must by this, his own assessment, be regarded as an inveterate (if not compulsive) gambler. But a more appropriate description would be that of an explorer, an adventurer, or a dreamer, someone eager to be at a frontier—in his case, at the forefront of science, where novel applications can create an industry with benefits for mankind and where the risk of failure is ever present.

For Alex, science is the driving force. With peerless vision, foreseeing the role of a technology in the marketplace, and with remarkable scientific intuition and understanding for development

Alex Zaffaroni

of products, he inspires confidence in investors to provide the necessary capital and convinces scientists and other key people to join him in his ventures. Unlike many business leaders, Alex remains emotionally attached to principles, projects, products, and people.

I recall a meeting of ALZA's scientific advisory board at one of the perilous times when defaults on bank loans were imminent. Someone suggested a drug-delivery device that might have wide appeal and be highly profitable. Alex dismissed it instantly, because it appeared gimmicky and did not embody the fine control inherent in other ALZA therapeutic systems. Good science and good technology might not be sufficient to make the company profitable, but, for Alex, a profitable company must be based on good science and good technology and on their honest application.

To obtain the high quality of science and technology needed for successful products, Alex has honed the management practices of others and, when necessary, has invented new ones. To his extraordinary vision for applications of science and technology to industrial products, he adds a special talent for selecting and inspiring others to fulfill his dreams. To them, he leaves the day-to-day business and technical operations, but he regularly examines these operations for their adherence to principle and to determine the need for fresh ideas and course corrections. With these people, he forms lifelong professional and personal attachments.

His underlying philosophy is faith in the creativity of people, with extensive reliance on young scientists and young business executives. He encourages them to exchange ideas, to publish freely, and to exercise entrepreneurial drive within (or even outside) the company. A do-or-die commitment to succeed is crucial, but never at the cost of concern for the welfare of the individual. The logo of ALZA's twenty-fifth anniversary, in 1993, bore the motto "Creating Value Through Innovation." The word *values* populates his language even more than *leverage*. These values include dedication to scientific soundness and quality of product, managed and leavened with human decency.

Alex now recalls an incident from his college days, in which Abraham Tugentman, one of his intimate circle of friends, was asked by them to explain some absences. Anxious and embarrassed to be interrogated, Tugentman answered that he was Jewish and had been attending the synagogue on a Jewish holiday, but he felt as if he had just confessed to a crime. For Alex, to observe someone dear to him being forced to endure the discomfort of isolation and discrimination made a searing and lasting impression. Alex's humane concerns embrace a wide circle—an employee unable to cope with a job, a friend or relative stricken with illness, a stranger seeking advice. Support also extends to causes for which his generous benefactions are usually anonymous, with the exception of large gifts to Democratic presidential candidates, the disclosure of which earned him a high place on President Nixon's "enemies list."

I have been asked, by mutual friends, "Is Alex Jewish?" The question is warranted. Many of his close friends and business associates are Jewish. His dedication to Israel and his philanthropy for the Weizmann Institute of Science, on whose board of governors he has long served, are well known. How then to explain this anomolous behavior of someone born and raised in another faith and exposed to teachings and myths that promote antisemitism? Think of antisemitism and similar prejudices as social viruses, each buried and silent in a recess of the nervous system, ready to emerge, on some provocation, as a virulent contagion. Characteristic of this disease is the violence that the carriers inflict on others. The victims of such violence acquire a sensitivity to the subtlest signs of a carrier, as well as an awareness of those who seem utterly immune to the virus. Alex is one of those rare people with a heightened immunity to all prejudices, including the virus of antisemitism, and, beyond that, with the dedication to combat them.

CHAPTER 4

A Biotech-Driven Pharmaceutical Company

At DNAX, ammunition was running low, the troops were worried, and the outlook seemed grim. By the end of 1981, the start-up fund of $4 million raised by Alex Zaffaroni from Swiss bankers was nearly consumed. The eight young scientists who had staked their careers on the viability of this venture were quite concerned, and I was anxious, too. The mood for investment in biotechnology had turned bearish. I had fished with Alex for joint ventures in Japan, France, England, and at home—a dozen casts without even a serious nibble. Never before had I been turned down for research support, but Alex had. A few years earlier, he had counted more than forty attempts to save ALZA from the brink of insolvency; all were fruitless until a deal was struck with Ciba-Geigy.

Alex had bought a little time for DNAX with a $1.5 million loan, at a high rate of interest, from Sanofi, the large French pharmaceutical company, which they hoped to convert to an investment in DNAX. Meanwhile, Alex kept casting about. He appeared assured when he walked into the office of Robert P. Luciano, the newly appointed CEO of Schering-Plough, to propose a joint venture with DNAX. Because it was Alex, respected for his achievements at Syntex and ALZA, Luciano listened closely, as did Hugh D'Andrade, who had represented Ciba-Geigy on ALZA's board of directors and was now responsible for strategic planning at Schering-Plough. A prompt response was essential, Alex explained, because DNAX needed cash to remain viable, and arrangements were being sought and made with other parties.

When Luciano, a lawyer and manager, joined Schering-Plough in 1978, he found a faltering company with desolate research resources for future products. His vision and courage in turning the company sharply into a biotechnology-driven enterprise needs to be seen against the background of prevailing practices in the pharmaceutical industry and the enormous gamble of investing in a novel technology with no proven products.

Origins of Schering-Plough Corporation

The origins and rivalries of large pharmaceutical companies bear remarkable resemblances to the feuding autocratic monarchies of nineteenth-century Europe. In the middle of that century, Dr. Ernst Schering, a pharmacist, began making and selling pharmaceuticals in Germany for a variety of common ailments, and, by 1880, he had entered the U.S. market. The business, with a reputation for high-quality products, prospered until the end of World War I, when it was taken over by the U.S. Alien Property Custodian and allowed to languish.

In 1928, Schering Corporation resumed American operations and began doing a thriving business with a widely advertised bulk laxative called Saraka, and, later, with Progynon, the hormone estradiol. In 1934, facilities were purchased in Bloomfield, New Jersey (later the corporate research center until 1993), and a production plant was soon installed in nearby Union. By 1938, the first commercial-scale production of testosterone brought the company's sales of prescription drugs to a level exceeding that of Saraka, and Schering became primarily a pharmaceutical company.

The current Schering-Plough emerged from the spoils of the last war with Germany. In 1942, the American branch of the German-owned Schering Corporation was acquired by the government of the United States as an alien property and sold at auction ten years later to an investment group headed by Merrill Lynch, which then took it public. In 1971, a merger with Plough, Inc., a Memphis-based consumer health-care products company, created the Schering-Plough Corporation.

During the decade after 1942, when Schering-Plough was managed by the Alien Property Custodian, it was fortunate to have had as its "temporary" guardian, and then as its president, Francis C. Brown, a young government attorney, and, as its vice president for research and development, Robert E. Waterman, a respected chemist who had shared in the discovery of vitamin B_1. Together, they implemented a "growth through research" strategy by investing in pharmaceutical research and recruiting excellent scientists.

A key start was made with the appointment of Erwin Schwenk, a talented steroid chemist and a long-time employee of Schering in Germany, who had relocated to the United States in the late 1930s as a refugee from the Nazi terror. Schwenk assembled an exceptional team of medicinal chemists who went on to make important new drug discoveries. Between 1948 and 1960, this team discovered pheniramine (Trimeton) and chlorpheniramine (Chlor-Trimeton), both antihistamines, perphenazine (a major tranquilizer), trichlormethazide (a diuretic), and diazoxide (an intravenous antihypertensive). Chlor-Trimeton alone and Chlor-Trimeton combined with aspirin (Coricidin) were both eventually approved for over-the-counter (OTC) sale, and both have enjoyed a worldwide market to this day.

To head the steroid group, E. B. Hershberg was recruited, in 1945, from Harvard, where he had been a long-time associate of the legendary Louis Fieser. In 1949, the dramatic appearance of cortisone as a treatment for arthritis and inflammatory disorders emboldened Schering to take a daring plunge into the intense competition to produce the drug. The complexity of its manufacture and the eventual decline in its price made cortisone unprofitable, but the effort yielded other benefits: It led Schering to the discovery of prednisone and prednisolone, anti-inflammatory corticosteroids far more potent than both cortisone and hydrocortisone and with less tendency to cause salt retention, the principal cortisone side effect. These drugs, known as Meticorten and Meticortelone, made Schering the leader in the corticosteroid market in 1955. Licenses were granted to foreign companies to make and sell these products in Europe and Japan; after complex litigation with American companies (including Upjohn, Merck, Syntex, and others), licenses were also offered for distribution in the United States. For four decades, these steroids have been the benchmarks of systemic corticosteroid therapy, as effective and affordable as any of the numerous steroids introduced later on. Subsequent introductions of betamethasone and beclomethasone and their various esters, developed jointly with Glaxo, helped Schering maintain a significant position in topical anti-inflammatory therapy.

A major Schering breakthrough came in 1960 through the combined efforts of George M. Luedemann, a mycologist in the fermentation department, and Marvin Weinstein, a microbiologist in charge of antibiotic screening. They discovered Garamycin (gentamicin), a multicomponent mixture of closely related aminoglycosides that is uniquely effective for treating *Pseudomonas* and *Proteus* infections, as well as infections by penicillin-resistant staphylococci. Luedemann had a friend, Americo Woyciesjes, an amateur microbiologist in Syracuse, New York, who owned a collection of actinobacteria of the genus *Micromonospora* that he had isolated from lake muds. Unlike their relatives in the genus *Streptomyces*, a pliable and popular source of antibiotics, the various strains of *Micromonospora* are finicky, but they became a source of unique antibiotic potencies, once various problems of cultivation had been solved. Broths from a particular strain were successfully fractionated by Schering chemists, but forecasted annual sales of only $1 million left management wary of further development. It was the energetic advocacy by Weinstein that brought Garamycin to market. As a broad-spectrum aminoglycoside for topical and systemic use, this antibiotic, during its patent lifetime, proved to be the most profitable product that Schering had ever produced.

Following Waterman's retirement in 1960 and Brown's accidental death in 1966, Schering was once again fortunate in the succession to leadership, this time by W. H. Conzen. He had been in the Schering organization in Germany, and he had found an opportunity, at the height of the Nazi frenzy, to transfer to the South African branch of the company, where he spent the war years. With the end of the war, he came to the Schering operation in the United States, and, by dint of astute judgments and political skills, became President and CEO. An austere man with a Teutonic manner, Conzen's smooth accession to replace Brown spared a bitter contest that might have fragmented the company. Conzen made some poor personnel choices, but his leadership was remarkable for two bold decisions that few who knew him would have predicted, given his conservative outlook. One was the merger with Plough, Inc., in 1971, signalling the entry of a purely pharmaceutical company into the very different business of consumer products. The other was the selection of Robert Luciano, in 1978, as his heir apparent.

Plough, Inc., was started in 1908 by 16-year-old Abe Plough, who peddled remedies by horse and wagon to farms and small towns around his native Memphis. A gifted salesman, an astute businessman, and a tough negotiator, he was able to expand his tiny company, through product development and acquisitions, into a leading manufacturer of health and consumer products. The line included such popular brands as Di-Gel antacids, St. Joseph aspi-

Ernst Schering

rin, Aspergum for sore-throat pain, Feen-A-Mint laxative, May-belline cosmetics, and Coppertone suntan lotions. Having reached the age of 79, and being happily engaged as Memphis's leading phi-lanthropist, Plough was persuaded by Conzen to merge his busi-ness with Schering and to assume the honorific title of Chairman of the Board of Directors of Schering-Plough. The combined prod-uct lines ranged from prescription and OTC pharmaceuticals to veterinary products, cosmetics, proprietary drugs and toiletries, a variety of home products, and even several profitable radio stations.

Abe Plough may have been the first Jewish chairman of the board of a major pharmaceutical company. Today, the staff and top managers in pharmaceutical companies and biotech ventures might find it difficult to imagine that the chemical and pharma-ceutical industries were virtually closed to Jews before World War

The young Abe Plough

W. H. Conzen (left) and Abe Plough after the merger of Shering and Plough

II. The speed with which companies lowered those barriers in the postwar period varied; Schering was among the swiftest. Key chemists who joined Schering after the war included E. B. Hershberg, Hershel Herzog, and Nathan Sperber. In subsequent years, Marvin Weinstein, Irving Tabachnik, and Preston Perlman, all of them biologists, and Frank Roth, a pharmacologist, held top-level research positions. Both physically and ethnically handicapped, Sperber, the coinventor (with Dominick Papa) of Chlor-Trimeton, confided to a colleague that he came to Schering because Jews couldn't get jobs at the more prestigious pharmaceutical companies. Richard Kogan, who is now Schering-Plough's president and chief operating officer, will become CEO when Luciano retires in 1995. The lack of prejudice at Schering was also evident in the employment of Italian and African-American scientists long before the current affirmative action and similar programs were in existence.

After the merger with Plough, the company grew from sales of $402 million in 1971 to $1 billion in 1978; it reached $1.74 billion in 1980, and $4.3 billion in 1993. A new microbiological and pharmaceutical research laboratory was opened in Bloomfield in 1974, employing 1200 scientists and technical staff. Alexander Z. Lane came from Bristol-Myers in 1977 to be the senior vice president in charge of research, and, in 1980, he became the president of pharmaceutical research. The Dr. Scholl's foot-care line (along with a large chain of stores in Europe) was acquired in 1979, and Wesley-Jessen, a contact-lens company, was brought in a year later.

The second bold decision by Conzen came in 1978, when the board expressed its concern about his replacement. "Don't worry. I have my eye on the right man." He identified Robert (Bob) Luciano, a 46-year-old lawyer, then at Lederle and formerly at Ciba-Geigy, whom he had met in meetings of the board of the Pharmaceutical Manufacturers' Association. Luciano agreed to come to Schering-Plough as the senior vice president for administration. The path to succeed Conzen was open, but Luciano had no assurances that the job would be his. After an interim appointment of Richard Bennett as CEO between 1978 and 1982, Luciano assumed the title and the leadership of Schering-Plough.

Soon after his arrival in 1978, Luciano heard, for the first time, of the new excitement about biotechnology from a presentation by Douglas Lawrason, who was then the vice president in charge of research and development. Luciano was intrigued. When he

In the 1920s, clever promotion was used to sell over-the-counter drugs, such as Feen-a-mint and Aspergum (both of which were later acquired by Abe Plough's growing company)

learned of Biogen's urgent need for capital, he recommended to Conzen that Schering-Plough invest $8 million in the fledgling company. Conzen readily agreed, recognizing that this percentage of Schering-Plough's cash holdings at the time was a gamble that he or Bob, on a strong hunch, would make with a comparable fraction of their personal assets.

Yet they both knew that the $8 million was only a small down payment for what would be needed for the development, production, testing, and approval of an utterly novel biological. By 1986, when the FDA approved of the first use of alpha interferon in hairy-cell leukemia, Schering-Plough had spent enormous sums in development of this recombinant-DNA product, more than twenty times its initial investment. In 1982, Luciano gambled again, this time with an investment of nearly $30 million to support basic research in immunology at DNAX, knowing—and even stating—that it would take ten years for any of its potential discoveries to reach the market, and at a cost many times the initial outlay.

The decision by this medium-size pharmaceutical and consumer-products company to be one of the first to plunge into biotechnology was not made in the board room or by one of the numerous committees of middle- and high-level managers. In the pyramidal structure of Schering-Plough, the direction came from the person at the apex, but with the advice (not needing the consent) of those who served him.

Robert P. Luciano

At the 1985 ceremony celebrating the opening of the new quarters for the DNAX laboratories in Palo Alto, Bob Luciano expressed his company's commitment to research as the foundation of its future. "Schering-Plough is not in business to do research. It is in research to do business. But science and business can, with some understanding, be viewed as two sides of the same coin. With such a relationship, we can bridge the gap between science and business, between discovery and application." What, in Luciano's training and experience, made him guide a third-rate, overly diversified pharmaceutical company with declining profits to become, within ten years, one of the most forward-looking and profitable scientific enterprises in the industry?

Robert Luciano

Born in 1932, Robert Luciano was the adopted child of a blue-collar family in a poor Bronx neighborhood. The love and support of his parents and his life-long devotion to them provide a happy contrast to the common frustrations of adoptive parents and to the need that many adopted children feel to trace their biological parents. As one of the bright kids in grade school, he was directed to the academically oriented Christopher Columbus High School in the Bronx, where he earned the good grades that qualified him to enter the tuition-free City College of New York, the only college he could afford to attend. Nobody in his family had ever attended college. He chose the accountancy curriculum at the downtown branch because it offered the clearest route to a white-collar job. Unlike the business courses, which he learned to hate, he enjoyed the required science courses. His disillusionment with business courses and curricula has not changed to this day.

Inducted into the Army upon graduation in 1954, he earned top place in his basic-training tests and was given the option of an assignment to be a clerk in military intelligence at the Pentagon. Despite detesting the regimentation and bureaucracy in his two-year stint, he came away believing in the overall value of universal military service. With the G.I. Bill of Rights to finance further education, Luciano decided to go on to law school. Columbia and Michigan, both outstanding schools, admitted him, and each offered a summer session that would permit an early start. Reluctant to return to New York, he chose Ann Arbor.

Law school was intellectual fun. Luciano found the courses interesting and the thought processes challenging. He made the Law Review and was in a position to compete for the best jobs. His prime concern was for a well-paying job to provide relief for his wife Barbara, who had been working to support them for the five years since their marriage.

Top law firms in the major cities paid $6500 a year, and he chose Rogers and Wells in New York. His accountancy background was put to use in tax work involving mergers, acquisitions, and pension plans of clients, one of whom was Warner-Lambert, the pharmaceutical company. Eight years later, Luciano was earning nearly $16,000 and was eager for a change of scene and a better salary. He had set his sights on $21,000. He would have taken a job with the Mattel Toy Company in Los

Robert P. Luciano

Angeles, but they would offer only $20,600. Both Upjohn in Kalamazoo, Michigan, and the Ciba Corporation (later merged with Geigy in 1970), in Summit, New Jersey, had jobs for him. The latter met his goal of $21,000. The pharmaceutical industry, with its intellectual appeal and social rewards, was attractive, but what mattered most, in picking Ciba, was the salary. It proved to be a wise choice, because he also found the people there to be a cut above those in other industries.

Richard Barth, then the general counsel of Ciba (and now its CEO), hired Luciano, and he, in turn, hired Hugh D'Andrade (who would, thirteen years later, join him at Schering-Plough). These three young lawyers might well have been one of the best teams ever in the pharmaceutical business. Luciano's new responsibilities required him to learn on the job, and his extensive experience in tax matters was of little use. In getting acquainted with the chemical, pharmacological, and clinical facts needed for maneuvering through regulatory agencies and marketing problems, he became aware of the flaws and limitations of the specialists who briefed him. Luciano and D'Andrade were diligent in exposing the dross and chaff in trying to get to the essence of an issue.

Luciano's nine years at Ciba-Geigy were marked by change and excitement, as he moved out of the law department into executive positions in marketing and pharmaceutical operations and absorbed the variety of challenges and modes of activities of these and related aspects of the industry at home and abroad. The cautious and deliberate pace of the close-knit Swiss management in Basel discouraged bold initiatives and expansion. The small U.S. branch of the Swiss corporation was so dominated by its foreign owners and directors as to leave little hope that an American could ascend to the higher reaches of company authority.

Raised in a poor family, Bob Luciano had managed, through academic talent and enterprise, to obtain a law degree and progressive advancement in the managerial ranks of pharmaceutical companies. In 1978, at age 45, when offered a position as President of the Lederle Division and Executive Vice President of the American Cyanamid Corporation, Luciano saw an escape from the sense of confinement he had experienced in working for Ciba-Geigy, as well as from its limited opportunities for growth. He had hardly settled in at Lederle when W. H. Conzen, the Schering-Plough

CEO, came looking for someone to groom as his successor. He had met Luciano when they were directors on the board of the Pharmaceutical Manufacturers' Association, where Conzen was the father figure and exercised major power. Conzen to Luciano: "Come over to us as Executive Vice President of Administration, with a seat on the board. If things work out, you have a good shot to succeed me." No promises were made. There were also two in-house candidates—Lee Jenkins and Frank Gleason—and things would simply have to sort themselves out. Luciano gladly accepted the offer.

When Bob Luciano arrived in 1978, the prospects at Schering-Plough were rather dismal. Garamycin, the flagship antibiotic, was about to go off patent. With far cheaper generic versions available, sales would plummet. No other marketable aminoglycoside was in sight. Schering-Plough's leadership position in steroids had been eroded by intense competition. Research in eight different therapeutic areas was being pursued by underqualified staff working with inadequate budgets. Luciano was convinced that research is where you succeed or fail in the pharmaceutical business, and that something radical had to be done to replace the failing effort.

Some weeks after arriving, he heard Doug Lawrason describe approaches made possible with the newly discovered recombinant-DNA techniques. Hormones and key body proteins, previously inaccessible for clinical use because they are found only in minute quantities in natural sources, could now be produced in massive amounts in bacterial factories. Several small ventures aiming to capitalize on these new techniques—Genentech, Cetus, Biogen— had already been started. But how was a big pharmaceutical company to enter this novel field, especially one so ill-prepared in the techniques and approaches of molecular biology?

A catalyst appeared in the person of Moshe Alafi, a Jewish immigrant from Iraq, settled in Berkeley, who was a shrewd investor in new ventures. He was a founder of Cetus and Biogen (and later of Amgen). Cetus was then promoting a machine for the automated recognition of bacterial colonies, and they had secured Schering-Plough as a client in the search for colonies that were superior producers of antibiotics. Alafi, acquainted with Schering-Plough managers and aware of Biogen's desperate need for cash, saw an opportunity for a Schering-Plough investment. Luciano was captivated by the possibility. Overcoming the indifference of his own research managers, he easily convinced Conzen to enter biotechnology by investing in Biogen.

With an infusion of $8 million, Schering saved the teetering Biogen. Schering acquired shares in the company, placed Luciano on Biogen's board, and secured rights to the products of three Biogen projects of the several that had been farmed out to collaborative university laboratories. Projects to produce leukocyte alpha interferon, fibroblast beta interferon, and erythropoietin were available, and all three had already been identified as targets by other biotech ventures. The markets were unknown—a grab bag of potential drugs in search of a common serious disease. Luciano's prime choice was alpha interferon, simply because he liked Charles Weissmann, who was directing the project at the University of Zurich. What matters most to Luciano is the person: "You bet on a horse because the jockey is right and the concept has intellectual substance." Schering-Plough's interest in the two other Biogen projects—beta interferon and erythropoietin—faded as the Biogen efforts were overtaken by competitive groups.

During those early Biogen years, from 1978 through 1981, Luciano attended the numerous meetings of Biogen's board of scientific advisors as well as those of its board of directors. He could see that the organization was flawed: Each of the founding scientific advisors—Walter Gilbert, Phillip Sharp, Charles Weissmann, Kenneth Murray, Peter-Hans Hofschneider, Heinz Schaller, and Brian Hartley—were outstanding scientists, but tensions had developed between the American scientists in Cambridge, Massachusetts, and the European scientists in Geneva. Much more profound were the cultural conflicts between the scientists and the investors (mostly Americans). The scientists regarded the business people as shallow mercenaries, and the business people looked upon the scientists as a necessary evil. Distrust among the disparate, collaborative science groups left each in doubt that the others would fulfill their assigned missions. The lack of collegiality and healthy give-and-take in scientific argument were heightened by the abrasive manner in which Wally Gilbert chaired the board meetings.

Charles Weissmann, with the invaluable help of a gifted postdoctoral fellow, Shigekazu Nagata (who had been trained by Yoshito Kaziro at Tokyo University), was the first in the race to clone alpha interferon. Weissmann delightfully recounts, in "The Cloning of Interferon and Other Mistakes," how the many meanderings in the research culminated in a highly controversial press conference, in 1980; announcement of the successful outcome drove up Schering-Plough stock by 8½ points. The investment of $8 million by Schering-Plough was only a small installment, for, by 1986, more than $170 million had been spent to develop large-scale procedures for a uniform, high-quality product of low toxicity that would meet FDA standards for the first clinical use.

The Schering-Plough microbiologists were seasoned in the large-scale production of antibiotics. They managed fermentations in 9000-gallon vats, in which well-behaved soil bacteria spewed out their rugged antibiotics, presumably as they do in their ecologic niches in the soil to mark a territorial prerogative in competition with other soil microorganisms. Purification of antibiotics from artificial growth media had become a standard procedure.

However, the production of a recombinant-DNA product from an engineered bacterium and the purification of a very unstable protein were beset by a host of novel technical and regulatory problems. First, one had to learn how to persuade a bacterial host to do something utterly strange—to produce a foreign protein, such as alpha interferon, in a monstrously large quantity. Then, the intricately folded protein molecule, roughly one hundred times the size of an antibiotic molecule, had to be extracted intact, uncontaminated even by traces of the toxins that are naturally produced by the bacterium. Beyond that, the fears of unknown dangers possibly attendant upon the use of recombinant DNA in medicine or agriculture required extraordinary precautions for the containment and disposal of the bacterial factories and myriad tests for assessing the biological activities of the product and those of any impurities that might be present in the final preparations.

Progress was slow in getting the cloned interferon gene into a suitable bacterial host that would tolerate it and express it sufficiently to yield significant quantities of the protein. Then there were delays in learning how to get the interferon out of disrupted cells without injury to the molecule. Finally, isolating the interferon molecule from the thousands of other kinds of protein molecules was quite another matter. After what seemed to Bob Luciano to be interminable confusion, the Schering-Plough scientists finally decided that they could proceed on a scale of 3- to 5-gallon batches, but they would not venture further. Simple calculations showed that the cost of a plant with thousands of such small production units would be prohibitive. Luciano's consternation was exacerbated by daily phone calls from Wally Gilbert berating the ineptitude of the Schering-Plough staff and the inadequacy of the company's operations.

At this juncture, Luciano, with only the most rudimentary knowledge of bacteria, proteins, or DNA, took charge of the interferon project; but, when Hugh D'Andrade joined Schering-Plough in 1981, Luciano transferred the direction of the project to him. As chairman of the team, Hugh had the invaluable assistance of J. Allan (Al) Waitz, a microbiologist, and T. L. (Nag) Nagabhushan, an organic chemist. Waitz, with ten years of experience in the isolation and characterization of antibiotics in another division of the company, was the project coordinator, and he was charged with keeping track of all the research aspects of the project. With his broad knowledge and no-nonsense attitude, Waitz tackled the large-vat cultures and used a team of his choice from throughout the organization to solve the obvious problems and the more numerous unanticipated ones. Nagabhushan, Director of Medicinal and Protein Chemistry, led the project's chemistry efforts, improving the recombinant-DNA constructs, perfecting the isolation of alpha interferon, and improving its yield manyfold.

Luciano remained concerned with the project, and he kept fretting over the colicky symptoms of his new baby. On one such occasion, Waitz had to say to Luciano, "It would help if you would get off the phone and let me do my job." With that, Luciano felt reassured that the project was in the right hands. Clinical trials were started within a year and a half, an astonishing pace of preclinical pharmaceutical development when compared with the norm of three to four years.

Despite repeated disappointments in treating selected types of cancer patients, a commitment had been made to spend $100 million to build a manufacturing plant and to accelerate expensive clinical trials for still other types of cancer. The first positive responses to alpha interferon had been reported in October 1983 in patients with an advanced form of hairy-cell leukemia, a rare cancer, which affected fewer than one thousand people. But with greatly broadened clinical studies undertaken in the steadfast belief that the drug would prove effective in treating more prevalent diseases, the number of cancer uses was increased and the remarkable effectiveness of alpha interferon against chronic hepatitis was demonstrated.

By the end of 1993, Schering-Plough's worldwide sales of alpha interferon (Intron A) for the year had reached $572 million, making it the largest-selling product in a single year in the company's history. Intron A is now marketed in 68 countries for 16 indications, including antiviral indications (against type B, type C, and delta hepatitis) and anticancer indications (against leukemias and lymphomas, for example). The wisdom of Luciano's scientific and business decision to enter biotechnology, and his entrepreneurial spirit and tenacity in the face of formidable odds, have been thoroughly vindicated.

Even before the interferon project, Schering-Plough's international operating unit had been attracted by financial inducements offered by the French government to invest in a research laboratory in France. But where to put it? The initial choices were Paris, Marseille, or Montpellier. At the prestigious University of Paris, the proposed Schering-Plough laboratory was torpedoed by a violent protest. Intrusion by an American company was caricatured as an American glutton consuming French brains. With government pressure to decentralize research and industry away from Paris, inducements were offered to select the Dardilly suburb of Lyon.

What was the laboratory to do? For lack of any sane and considered planning, Preston Perlman, head of research operations and a francophile with a love of good food and wine, quixotically chose immunology as the subject. After all, he reasoned, France was acknowledged as the birthplace of immunology. The apotheosis of Louis Pasteur as the father of immunology still prevailed a century later. But what research was Schering-Plough to do in immunology?

It was not surprising that Bob Luciano, having assumed the commitment to develop interferon and now saddled with a French connection to immunology, would, in 1979, say to Frank Bullock (who had just arrived to succeed Perlman): "What is this immunology bullshit?" Whereupon Bullock assembled a task force of academic immunologists: Harvey Cantor from the Harvard Medical School (Dana-Farber Cancer Institute), Irving Weissmann from the Stanford Medical School, Leonard Chess of the Columbia University College of Physicians and Surgeons, and Richard Gershon of Yale.

The lengthy report of the task force advised Schering-Plough management (not surprisingly) that immunology was a field in which the company might wisely invest its resources and that the T cells of the immune system, an area of special interest to the members of the task force, should be the company's primary focus. T cells were judged to be the most important agents in modulating the body's immune responses in health and disease, and research on T cells would nicely complement Schering-Plough's earlier decisions to concentrate on inflammation, allergy, and infectious disease.

Any doubts that Conzen and the Schering-Plough board might have had about Luciano's succession to the leadership of the company were rapidly erased by his vigorous management style and his energetic commitment to strengthen the company's enfeebled research foundations for drug discovery. Entry into biotechnology was clearly a bold investment for the future, but more needed to be done. How could he implement the recommendations of the immunology task force to work in this unfamiliar and amorphous area? None of the established people on the task force would consider a move to an industrial setting, least of all to one as mediocre as Schering-Plough. Nor could a significant number of bright young people be attracted to choose the feeble science and drab geography of Bloomfield or the nondescript laboratory in France.

At this moment, Alex Zaffaroni entered Bob Luciano's office. Luciano had known Alex from an encounter some eight years earlier, when Alex had negotiated a Ciba-Geigy deal with ALZA. Hugh D'Andrade knew Alex even better because he had been the Ciba-Geigy representative on the ALZA board. Both had the highest regard for Alex as an extraordinary scientist-entrepreneur with vision, charm, and integrity, but their colleagues would have to be persuaded.

Alex described DNAX, only a year old, led by a star-studded scientific advisory committee, but with fewer than a dozen young scientists just beginning to work in a remodeled ALZA laboratory in Palo Alto. Their objectives were to design genetically engineered antibodies for use in drug overdosage, diagnosis, and industrial procedures. Short of money, DNAX was seeking research funding in exchange for some equity in the company and access to some of its discoveries.

By this time, Schering-Plough's relationship with Biogen had curdled (it had become a love–hate affair with little affection or trust remaining), thus creating a clear ground for a new and significant Schering-Plough venture in biotechnology. There were hurdles to be overcome at several levels at Schering-Plough and at DNAX—attitudes, people, and money—but Luciano believed that the company's future, as with all the pharmaceutical industry, was in biotechnology-driven biopharmaceuticals. He believed that it was tactically right to be associated with DNAX and strategically best to own it.

A deal was struck, in May 1982, that pleased both sides and became sweeter as the years went by. Of course, there were doubters and detractors. With Garamycin off patent, Schering-Plough profits had plummeted 25 percent in 1981, and the price for DNAX was high. DNAX had no assets and only eight young scientists, and one Wall Street analyst derided Luciano for having paid more than $3 million per Ph.D. Moreover, an intensive trial of alpha interferon on cancer patients that had been reported at the annual meeting of the American Society of Clinical Oncology was disappointing; only a small percentage of the patients showed a partial decrease in tumor size, and unanticipated toxicity problems had surfaced. The word spread around the industry and around Wall Street that "interferon looks like a loser" and that "unlucky Luciano has lost big in his biotech crapshoot."

If it can be said that Alex Zaffaroni fathered DNAX, then surely Bob Luciano can be credited for adopting the infant and devoting himself to its nurture and maturation. Such is the absolute authority vested in the CEO of a corporation that a gifted, resolute leader can dramatically transform an organization. I recall an incident in which a group of us, including Luciano, were discussing the news that David Baltimore had finally been appointed President of Rockefeller University over much-publicized objections by a number of dissident professors. "Well, now he can go in and clean house," said Bob. To which I had to respond, "Did you ever hear of tenure? He'd have trouble firing the janitor."

As Chairman of the Board and CEO of Schering-Plough, Bob Luciano answers only to the directors on his board, highly worthy people from many walks of life and varieties of business. They enjoy his good humor and respect his remarkable vision in directing Schering-Plough to a biotechnology-driven future. Above all, they are aware of his managerial skills, which achieved 20 percent annual growth in earnings per share from 1986 to 1992 and 18 percent in 1993—unexcelled in the industry. The only time he was overruled by the board was in his request to extend the board memberships of Milton Rosenthal, Harold Berry, and W. H. Conzen when they had reached the mandatory retirement age of 72. Beyond an affection for them, he had come to rely on their advice and judgment.

The board of Schering-Plough, unlike the boards of other pharmaceutical companies, has not included scientists who have any particular knowledge of, or association with, biology, chemistry, or medicine. Having selected his research and professional managers, Luciano does not want to see their tactics or his overall strategies questioned or closely examined. With his vigorous expansion of sound science and the effective development of discoveries into profitable drugs, there is little reason for the stockholders or their board representatives to complain.

Luciano has also groomed some outstanding managers. To forestall their being tempted by higher-level jobs in other companies, he announced his intention to retire, in 1995, at age 63, so that Richard Kogan could succeed him as CEO, thereby creating a succession of openings for promotions in the top managerial ranks.

Hugh D'Andrade

Hugh D'Andrade, Executive Vice President in charge of administration at Schering-Plough, is also responsible for strategic planning. He played the key role in the negotiations to acquire DNAX, and has served ever since as the chairman of its policy board. More than anyone else, D'Andrade has managed to reconcile seemingly disparate interests: those of the DNAX scientists, to acquire knowledge for its own sake, and those of Schering-Plough managers, to make the research promote its business. He has been unstinting in his attention, devoting whatever time and effort it required to adapt procedures and policies for the success of this utterly novel enterprise.

Hugh D'Andrade

D'Andrade intended to be a lawyer when he entered Rutgers University, and he did go on to get his LL.B., with honors, from Columbia University Law School in 1964. He had early interests in sociology, influenced both by an older brother, who held a professorial appointment in cultural anthropology, and by his Christian Science upbringing. In college, his faith began to waver; his dabblings in hypnosis were clearly antithetical to Christian Science. Never having been to a medical doctor before his twenties, he then entered a career of hobnobbing with them and dispensing their intrusive drugs.

Instead of electing biology to fulfill the college science requirement, he chose math and physics over the more descrip-

Hugh D'Andrade

tive science. With a keen, analytic mind, and stimulated by outstanding legal scholars at top-ranked Columbia, D'Andrade
might have been expected to lean, as did many of his classmates, toward academia or a high-powered Wall Street law firm.
Yet, his initial intention was to settle into a general law practice
in the small New Jersey town of Metuchen, in which he had
grown up.

A chance interview diverted D'Andrade from a small-town
practice to a year's clerkship with an associate justice of the New
Jersey Supreme Court, which, in turn, opened the door to the
medium-sized firm of Toner, Crowley, Woelper and Vanderbilt
in Newark. He found litigation frustrating and unfulfilling:
one was saddled with a set of facts and then was required to
rearrange them to suit the client's case. Endless briefs and
depositions, and the frustration of waiting for judges and
juries, preceded the trials that almost never came to pass.

In 1968, D'Andrade, after three boring years practicing corporate law, answered an ad for a job with the Ciba Corporation

in Summit, New Jersey, the American branch of the huge Swiss pharmaceutical company. Interviews with Richard Barth and his assistant, Bob Luciano, persuaded him to sign on, at age 30, as general attorney. Working for Luciano, five years his senior, D'Andrade was convinced that, within two years, he could take over Luciano's job as counsel to Ciba's pharmaceutical division. That is, in fact, what happened, but only because Luciano was promoted to Vice President of Marketing.

The Federal Food, Drug, and Cosmetic Act of 1962, authored by Senator Estes Kefauver, had been a bombshell exploding on the pharmaceutical industry. Drugs, new and old, had to be proven efficacious as well as safe. Panels of the National Research Council undertook a total review of the major drugs, and they recommended the recall of many. D'Andrade's job was to work with Ciba clinicians to handle these challenges, to check that claims for Ciba's new drugs were justified, and to ensure that the company's advertising did not exceed those claims. He learned the required chemistry, pharmacology, and medicine on the job by asking searching questions. These apparently naive queries, taken as intellectual arrogance, nettled some of the clinicians, who angrily told him that they didn't have the time to give him a medical education.

Even so, D'Andrade never felt intimidated. As long as FDA regulations and package inserts were in a language that approximated English, he believed himself able to understand them. As his interest in human biology deepened, along with his interest in chemical interventions to improve health, so did his facility to grasp problems and to penetrate to the basic issues. With no training in chemistry or patent law, he could still deal with technical, legal, and administrative problems as they arose in the laboratory and in the marketplace. Was Ciba correct in promoting a stimulatory drug for children? Was lifetime use of an antihypertensive drug really safe and justified? Was Ciba negligent with some of its many products, as claimed in suits for injury that had been brought against the company? These were all questions that D'Andrade, with his newfound knowledge of science, felt competent to handle.

The merger of Ciba and Geigy in 1970, equal both in size and in dedication to research, created a mammoth player in the industry. With this expansion, D'Andrade's responsibilities in the American operations increased. Initially, as Counsel of the Pharmaceuticals Division, he also became Vice President of Administration, putting him in charge of about 3000 people, union relations, business development, licensing and acquisitions, and problems ranging from research to marketing. When Ciba-Geigy acquired a controlling interest in ALZA, he served on its board of directors from 1978 to 1980. He came to know Alex Zaffaroni and, with that, gained a profound respect for his vision and integrity.

By this time, Luciano had moved to Schering-Plough, after a one-year interlude at Lederle, and was being groomed to take over as CEO. In early 1981, he invited D'Andrade to come to Schering-Plough to take his place as Vice President of Administration and to continue the activities he had performed at Ciba-Geigy—particularly, strategic planning of pharmaceutical business based on vigorous research. Luciano had D'Andrade replace him on Biogen's board of directors in the seat reserved for Schering-Plough. During his five years on the Biogen board, D'Andrade also attended the turbulent and exciting meetings of the scientific advisory board, which was composed of Biogen's founders, later joined by Jeremy Knowles and Walter Fiers and chaired by Walter Gilbert.

At Schering-Plough, D'Andrade also took over as the chairman of the interferon project team and guided it for nearly three years, until the manufacturing and marketing aspects became the dominant issues. When the phenomenal success of Intron A, with more than $500 million in sales in 1993, was heralded in a handsome brochure, D'Andrade credited Luciano for his extraordinary courage and vision in entering biotechnology, and he praised the members of the research and development teams for their many contributions. But there was no mention of the vital role he himself had played in the critical early stages of the operation.

In the same year in which he arrived at Schering-Plough, D'Andrade became the key figure in the negotiations with Zaffaroni for the purchase agreement with DNAX.

Acquisition of DNAX

The DNAX proposal to Schering-Plough came at an opportune time. The company was poised to inaugurate a program in immunology, had intended to sever its connection with Biogen, and was seeking an even stronger tie to biotechnology. Frank J. Gleason, Hugh D'Andrade, Alex Lane, J. Allan Waitz, Frank Bullock,

The DNAX building, 1985

and others from Schering-Plough made several visits to the DNAX laboratory to meet, and to exchange presentations with, the scientists, the Stanford founders, and the staff. These were followed by discussions Zaffaroni had, at Schering-Plough, with Luciano and D'Andrade. Interest mounted on both sides. The Schering-Plough managers recognized that motivated scientists and advisors of the quality found at DNAX would be impossible for them to assemble on their own. At issue was an arrangement that could assure DNAX of maintaining its focus on basic science and assure Schering-Plough of a return on a substantial investment.

On the DNAX side, relief from a precarious financial position and from the constant need to seek and meet contractual agreements was of prime importance. Along with DNAX's strong desire for secure, long-term support to pursue molecular biology came a willingness to apply these novel techniques to fundamental questions in immunology of interest to Schering-Plough. Neither Alex nor the scientists and founders had ambitions for sudden wealth, such as that created by the Genentech's initial public-offering bonanza, and there was no wish to build DNAX into a fully integrated pharmaceutical company. After all, the name DNAX Research Institute of Molecular and Cellular Biology had been chosen to attract scientists rather than investors.

On the Schering-Plough side, D'Andrade, in ten pages of handwritten notes dated 22 December 1981 (after a meeting with Zaffaroni, Gleason, Lane, and Harold Hiser), compared the merits of linkage to DNAX by means of a joint venture with the merits of linkage by total acquisition. Complete control, backed strongly by Luciano, was justified by D'Andrade in his notes as follows:

> The science represents a fundamental, new approach with potentially practical applications. Work in this field and with the DNAX Advisory Board could have beneficial fallout for the direction of the immunology project at Schering-Plough. Work with DNAX/ALZA could have benefits for interferon, such as new dosage forms. The DNAX approach with Zaffaroni's selling effort will have a beneficial effect on our stock value [and] taking a majority ownership will have the best optical effect re stock price. . . . the impact of the West Coast technology concentration would have good optics—bold move. Our Biogen relationship is deteriorating—we need [a] replacement. [The expected costs:] Buy out Swiss ($8 million), buy out ALZA ($5 million), buy out insiders ($10 million, payable 3/31/87 in Schering-Plough equivalents).

The cost would fit into projected Schering-Plough research spending of $126 million for 1982, a 16 percent increase over 1981. A compounded 15 percent annual increase in research spending for 1982–1987 would make Schering-Plough one of the ten leaders in the pharmaceutical industry, after Merck, Johnson & Johnson, Lilly, Pfizer, Upjohn, Smith-Kline, and Bristol-Meyers, and ahead of Abbott and Squibb. The costs would be offset in part by savings in the immunological laboratories in France and from a tax credit for increased research costs in the United States.

> We would put in CEO; Zaffaroni would stay on as Chairman. How to attract the best people? We must give them some independence and hope that what they do is still of interest to us. Requires leadership; I can say of my own knowledge that Zaffaroni is good at that. This opportunity window will close [that is, other investors and partners may soon complicate the picture].

Among the top managers, Bullock, under whose authority DNAX would be placed, had yet to express a firm opinion. Initially, an arrangement like the one with Biogen had been contemplated, but it became clear that, unlike Biogen's interferon, there was no DNAX project to which they wanted rights or that would soon become available; the most attractive thing about DNAX was its people.

Late one night, after a long day of meetings and discussions at DNAX, D'Andrade, alone with Bullock, asked him point blank: "Frank, tell me whether you want to have DNAX as part of your research operation. I don't want any analysis; no reasons. Just a yes or a no. If you say 'no,' we should terminate our discussions. If you say 'yes,' I will find a way to get it done. Now just tell me: yes or no?" Bullock, weighing many factors (not the least of them the relief of turning over the messy immunology operation Schering-Plough had undertaken in France to a future DNAX manager), hesitated. Then, typically, a man of very few words, he said: "Yes." What if he had said "No"? Bullock and others believe that another arrangement would have been found for acquiring DNAX.

In the ensuing weeks, a five-year strategic plan that focused on basic studies of T cells in the immune system was presented by Bullock and Lane; the DNAX scientists and founders responded with enthusiasm. The personal interactions were also excellent, except for the discordant note struck by William O'Neill, Vice President for Corporate Development and a principal figure in the DNAX organization. Bill, believing that Schering-Plough, based on its record, could not be trusted to let DNAX do basic research in a free and open atmosphere, expressed skepticism throughout the negotiations. The financial arrangements reached were much like D'Andrade's estimates, except that the Swiss bankers demanded $13.75 million, a return of more than 300 percent on a one-year investment. The total price, including the Schering-Plough stock distribution, which was to be spread over five years, came to $27 million.

In seeking approval of the Schering-Plough board for the acquisition of DNAX, Gleason could cite few tangible assets. The 25,000-square-foot laboratory, in space leased from ALZA, could house a research staff of 50 to 60, plus administrative personnel, and had cost $1.5 million to build and equip. "The principal assets," Gleason argued, "are the scientific advisory board and the

Alex Zaffaroni

Alex Zaffaroni and Frank Gleason signing the DNAX–Schering-Plough agreement in 1982

working scientists, recruited with the help of the advisory board and assembled by Dr. Zaffaroni."

The eight working scientists were described by Gleason as

> a first-class team . . . which could not be recruited to work in our, or a competitor's, industrial laboratory. These include former prize "postdocs" and students directed to DNAX by the scientific advisors. [They are] highly motivated by the professional opportunity they perceive in the DNAX setting and [by] attractive rewards. . . . Linking these capabilities with Schering-Plough's fully developed pharmaceutical industrial competence will put in place the resources required to achieve our strategic goal to be among the first—if not the first—to produce, through this advanced technology, new agents in major therapeutic markets. The U.S. market alone for antiallergens is at present $200 million; for antirheumatics, $180 million; and $1 billion for anti-infectives.

Other intangible DNAX assets included license agreements— one with ALZA for unique drug-delivery systems and one with the Massachusetts General Hospital for antibody products for cardiovascular diseases: Dig-Annul, for treatment of overdigitalized patients; Renen-Dart, for malignant hypertension; and Thrombo-Lys, for dissolving blood clots.

With presentations by Gleason and D'Andrade and a persuasive appearance by Zaffaroni, the board approved the purchase of DNAX on 29 April, 1982, and the sale was formally completed on 13 July. An announcement had been made a month earlier that Schering-Plough had selected J. Allan Waitz to be President and CEO, which had filled all of us at DNAX with trepidation. Waitz, a 16-year Schering-Plough veteran, had been a vice president in charge of microbiological research, and he had just coordinated the successful production of recombinant alpha interferon. Still, he was a company man without distinction in basic research, let alone in molecular biology or immunology. It seemed that Waitz had been installed as a "colonial governor," and Bill O'Neill's prediction that Schering-Plough would exert tight control over DNAX research appeared, to those of us at DNAX, to be prescient.

Nor, for that matter, was Allan all that pleased with the DNAX assignment. He was comfortable in his position and happy with his lifestyle in New Jersey, and he was nervous about the prospect of moving and about many other aspects of the DNAX arrangements.

The research goals had not been clearly defined, and each of the scientists had several projects under way—shades of the diffuseness of research at Biogen, which had so discouraged the Schering-Plough people. How could DNAX be directed, from a great distance, to focus its efforts on strategic Schering-Plough objectives? And how could DNAX technology and discoveries be transferred to Schering-Plough? The Dardilly laboratories in France would also become Waitz's responsibility, and they seemed unmanageable. It was, in Waitz's view, "a lose–lose situation." However, D'Andrade, with the foresight that Waitz was the right person for the job, was able to persuade him to take on the DNAX venture.

Our fears at DNAX about Waitz's appointment were allayed by several assurances from Schering-Plough: Waitz would serve as interim CEO for only one year, and we would then have a major role in the choice of his replacement. DNAX policies would be preserved. A new policy board of nine members would include the CEO and the four DNAX founders to balance the four top Schering-Plough managers (D'Andrade, Lane, Bullock, and Kogan). Finally, a committee of four, constituted to assess attainment of the annual performance goals and thus empowered to determine whether Schering-Plough stock should be released annually to the scientists and founders, would be made up of three founders (Berg, Yanofsky, and me) and a single (and sympathetic) member of the Schering-Plough board, Guy Stever, a physicist.

As events proved, the Waitz appointment was one of the best things to happen to DNAX.

J. Allan Waitz

Al Waitz, born in 1935 in Elizabeth, New Jersey, not far from the Schering-Plough plant in Union, was the first in his family to go to college. An avid fly fisherman and naturalist, he chose the University of Idaho, intending to study forestry and fisheries. Instead, he became interested in zoology and was inspired by a parasitologist with whom he stayed on for a master's degree. With fellowships from NIH (National Institutes of Health) and NSF (National Science Foundation), he moved on to the University of Illinois for a Ph.D. in biological sciences, and there he encountered top minds and vigorous science—Irwin Gunsalus in biochemistry, Sol Spiegelman in microbiology, and Ladd Prosser in comparative physiology.

J. Allan Waitz

His thesis, on the biochemical aspects of the tapeworm life cycle, opened job opportunities at the NIH, Kansas State University, and Parke-Davis. He chose the pharmaceutical company for the latitude and resources it provided for studies of host–parasite relationships; the offer included an advanced status, being second in command in a group of twenty. After three or so years, the erratic behavior of the department head made his position untenable.

A move to Schering-Plough, in 1966, had several attractions. There was excitement over the discovery of the potent aminoglycoside antibiotic Garamycin (gentamicin) and a need for its biological evaluation. There was also an intimacy within the organization that made lunch with Francis C. Brown, the CEO, not uncommon. Over the ensuing years, Waitz was placed in charge of departments devoted to more rational screening of aminoglycosides, the discovery of novel antibiotics, and studies of resistance-factor plasmids, which had just come into view. He also directed the fermentation group and handled a collaboration with Cetus for strain improvement, a competitive and frustrating partnership.

When Bob Luciano thrust Schering-Plough into biotechnology, with a commitment to develop a Biogen product, Waitz became the coordinator for the alpha interferon project. Even though Biogen, with Charles Weissmann's efforts, was the first to clone alpha interferon, the necessary improvements in vector design and host expression systems and large-scale isolation procedures were not being developed at a proper pace.

The Biogen–Schering-Plough collaboration was failing, partially because, at Biogen, the work was complicated by Biogen's ambitions to become a full-fledged pharmaceutical company, and, at Schering-Plough, there was little experience or talent in the new area of molecular biology. On top of this, communications between the two, an ocean apart, were confused by a lack of trust and a lack of candor; and the Roche–Genentech competition was moving ahead. At this juncture, Waitz joined the project, with Luciano's promise to provide all the financial and administrative support he would need for Schering-Plough to be "the first to file for regulatory approval for a real clinical use of alpha interferon in a meaningful market."

Resolute in the face of doubts expressed by the world-class Biogen scientists (the likes of whom he had not met since graduate school) and unfazed by Schering-Plough authorities (including Luciano), Waitz plunged ahead. He recalls a fateful meeting with senior management when he was finally able to announce, "I think we have a product." The interferon team had developed a stable formulation and had made adequate quantities of the substance, which had passed toxicology tests and had shown clear efficacy in early clinical trials.

In 1981, Waitz was named Vice President of Microbiological Research and put in charge of a staff of 150. He was made responsible for all aspects in the discovery and evaluation of novel natural and synthetic antibiotics and in the application of the new recombinant-DNA technology. In this capacity, Waitz was a member of the Schering-Plough team that visited DNAX to explore the various aspects of a possible connection. As President and CEO of DNAX during its first ten years under Schering-Plough ownership, Al Waitz was crucial in defending and explaining, to the huge pharmaceutical company in the East, the open, academic atmosphere of its tiny satellite in the West. Waitz had spent 20 years in the pharmaceutical industry, the last 16 with Schering-Plough. He knew the bureaucratic layers, the procedures, and how to use or circumvent them. He also understood and could encourage the unconventional research approaches and lifestyles at DNAX.

For their considerable investment in DNAX, Schering-Plough management wanted something tangible in discoveries for product development beyond the enhancement of their image as a biotechnology-oriented pharmaceutical company. They recognized that they needed to attract and retain a world-class scientific staff, as well as to sustain the allegiance of the DNAX founders and scientific advisors. To achieve this, Waitz nurtured an academic atmosphere that encouraged prompt publications; the sharing of reagents, cell lines, and techniques (outside DNAX as well as inside); unstinted physical resources; and generous perks and compensation.

Waitz guarded the status of DNAX as a legal entity separate from Schering-Plough. With authority for purchasing, personnel, payroll, safety, legal matters, maintenance, finance, and so on, Waitz kept the massive arms of the company bureaucracy at bay and could pioneer a day-care program, language instruction, open seminars, discounted refreshments, streamlined salary grades, work-filled corridors, minimal security regulations, short-term appointments for visiting scientists, and an ambitious postdoctoral program.

In formulating and implementing long-range plans, Waitz relied on the DNAX founders and Schering-Plough managers meeting quarterly as a policy board. These meetings also provided settings for building warm personal relationships and a sense of trust that cemented the disparate academic and industrial cultures. Monthly meetings of the founders with Waitz, constituted as an executive committee, contributed to scientific evaluations, recruiting, and some operational decisions. Creating and strengthening the DNAX–Schering-Plough connection depended on all concerned —Waitz, the Schering-Plough managers, the DNAX founders, and the dedication of the spirited scientists and staff.

CHAPTER 5

Growth of a Biotech Venture

Schering-Plough's acquisition of DNAX, and the consequent reorientation of the research focus to T cells, came about rather easily. The initial objective of engineering antibodies was not entrenched. During the six months or so of the laboratory's existence, only Kevin Moore had started work on antibodies. Donna Rennick and Tim Mosmann arrived late, and Mosmann was already familiar with T cells; the other five were outfitting their laboratories, training technicians, continuing their own projects, and setting up systems for cloning and expressing recombinant DNA.

The antibody programs, pursued only by Moore, lasted for just a year. A patent issued to Zaffaroni and Moore for "minimum binding compositions" is one of the remnants of the flagship of the initial DNAX expedition. The patent may have some value, in that vigorous programs of antibody engineering have since been undertaken by several industrial and academic laboratories. A more significant consequence of this early interest was the expert and efficient antibody-production program set up and directed by John Abrams, who joined DNAX early and provided an invaluable resource for all the subsequent cytokine work.

Whether DNAX could have accomplished its antibody missions a decade ago by producing marketable products will never be known. In the early solicitations to pharmaceutical companies and investors, I confided to Zaffaroni my belief that the antibody work at DNAX would be superseded, once research was under way. It seemed likely that bright young scientists, applying the novel cloning techniques at the forefront of a dynamic field, would be changing directions in response to discoveries made by them and others.

To assuage Schering-Plough's doubts about the DNAX antibody program, Zaffaroni expressed the same conviction—that the power of this new technology in the right hands would create opportunities even more fertile than those projected.

The Schering-Plough acquisition, with the attendant shift of focus away from antibodies, removed a potentially worrisome issue. The DNAX scientists had been recruited with the promise that they could devote a substantial fraction of their time to their own research interests. It was hoped that the work and ideas of those talented and congenial people would eventually shape a coherent program. But there was very little interest in antibodies, and there was a consequent danger that the wide diversity of research directions would persist and grow. The Schering-Plough initiative to explore the molecular biology of T cells and cytokines, adopted with uniform eagerness, brought the instant cohesion that would have taken a long time to achieve, if ever.

The Early Golden Years:
Dawn of the Cytokines

The cytokine "gold rush" of the 1980s, in which DNAX staked some major claims, was based on explorations of a decade earlier by Donald Metcalf in Melbourne, Australia. His work revealed the dependence of immune and blood-forming cells on certain protein factors for their growth, development, and function. The first such were identified as colony-stimulating factors (CSF). The factor for granulocytes was G-CSF, marketed in 1991 as Neupogen by Amgen; the factor for both granulocytes and macrophages was GM-CSF, launched in 1992 as Leucomax by Schering-Plough and in 1991 as Leukine by Immunex. Later it became apparent that these and other factors acted on several cell types, eliciting a variety of functions.

Names of factors initially chosen for a particular action on a single cell type became confusing. To designate a protein produced by one kind of white blood cell (leukocyte) that acted on other cell types, the more general term *interleukin* was adopted; numbers were assigned in the order of their discovery (IL-1, IL-2, IL-3, and so on). More than a dozen interleukins have been isolated and characterized, and their genes have been cloned. With the recognition that these hormone-like proteins operate on cells other than leukocytes (for example, erythropoietin acts on the precursors of red blood cells, and IL-3 acts on mast cells, the producers of histamine), the even more general term *cytokine* has come into use.

The initial success of DNAX rested on four elements:

- the decision, directed by Schering-Plough strategy, to work on T cells at an opportune time when recombinant-DNA techniques would have a major impact on immunology;

- the introduction at DNAX of a novel "expression" approach to the cloning of recombinant DNA;

- the intense drive of Kenichi Arai; and

- an ambience that promoted creative contributions to a communal effort.

The T cells, named for their development in the thymus gland, regulate the body's immune responses, in part by sending out cytokine signals to two kinds of cells. From B cells (originally discovered in the bursae of chickens and later in the bone marrow of mammals), the T-cell signal elicits a "humoral" response resulting in the release of highly specific antibodies into the blood and body fluids. The T-cell signal to certain tissue cells that engulf foreign material (such as bacteria, viruses, or parasites) evokes a different response that resembles tissue reactions to inflammation (Figures 7 and 8). How T-cell signals are tuned and balanced to evoke either the humoral response or the tissue-type response had been a major mystery in immunology. The solution, discovered at DNAX, was a fundamental contribution to basic science. It established the international scientific stature of DNAX, and it set the stage for the discovery of IL-10, one of the most important of the cytokines.

The expression-cloning technique advanced at DNAX became feasible by exploiting a method that had just been developed by Hiroto Okayama, a postdoctoral fellow with Paul Berg at Stanford. The Okayama–Berg procedure produced full complementary DNA copies (cDNA) of messenger RNA (mRNA), even in crude cell extracts containing thousands of different mRNAs. When these stretches of cDNA were recombined into the tiny DNA circles known as plasmids and introduced into an appropriate host cell, they functioned as genes to generate the proteins encoded by the mRNAs.

The first project undertaken at DNAX was to clone and identify a factor made by T cells that promoted the growth of mast cells, known to be responsible for the release of histamine and other allergic reactions. Many hundreds of pools of plasmids, each pool containing hundreds of different plasmids, were screened for the capacity to make a host cell, into which a single plasmid had entered, produce the mast-cell growth factor. From more than a hundred thousand plasmids, a single positive was found. Large-scale production and isolation of the factor from cells bearing this plasmid established its identity as IL-3, a growth factor that had been named previously but whose activity had been ascribed to a crude mixture.

Although the pure IL-3 supported the growth of mast-cell cultures, the final extent of their growth fell short of what could be obtained with crude IL-3 preparations. This observation led to a search for the missing component and the ultimate cloning and isolation of a new cytokine, named IL-4. With pure IL-4 in hand, the activities of growth factors in crude preparations that had been reported for B cells and T cells and for stimulation of production of immunoglobulin E (IgE), a prominent antibody of the allergic state (Figure 9), could, in each case, be ascribed to IL-4.

The immunologists at DNAX—Coffman, Mosmann, and Rennick—kept increasing the range and facility of cell-culture assays for new factors. The molecular biologists—the Arais, Lee, and Zurawski—sharpened the cloning and isolation procedures for new factors. With other young scientists who joined them, and with ancillary support for the preparation and chemical analysis of monoclonal antibodies, the DNAX team would, within three years, file patent claims for the discovery of mouse IL-3, IL-4, IL-5, and IL-6, for mouse GM-CSF, and also for some of the human versions of those cytokines. [IL-4 is now in exploratory (phase II) clinical trials for the treatment of cancer and life-threatening immunodeficiency diseases, with expectations at Schering-Plough that it will find important uses and prove to be profitable.] In the course of their efforts, the DNAX teams had assembled a formidable collection of virtually all the known cytokines, along with assays and antibodies that bound them—a unique resource for exploring the functions of these factors and for verifying the discovery of new ones. A major force propelling the DNAX effort in these early years was Kenichi Arai.

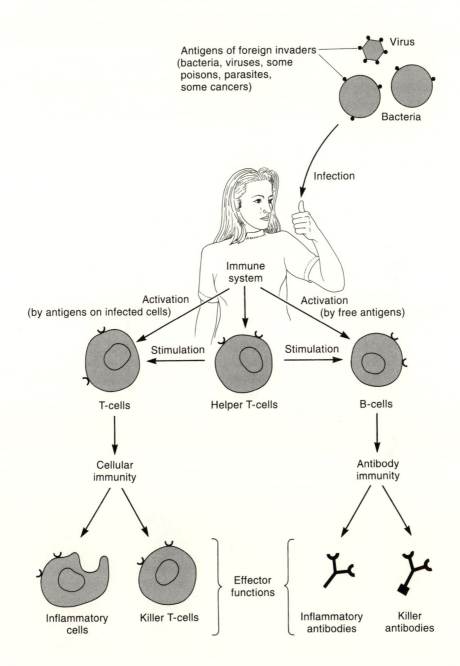

FIGURE 7. The immune system protects against infection by isolating and killing (lysing) foreign invaders.

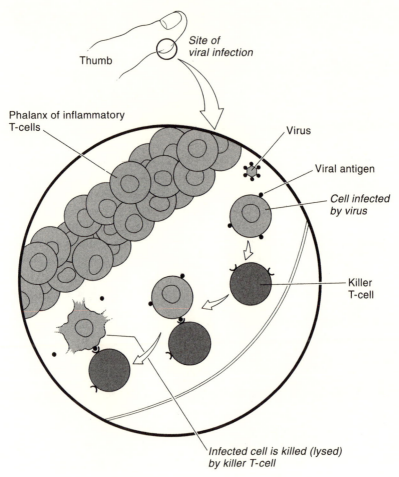

Cellular immunity

Thumb

Site of
viral infection

Phalanx of inflammatory
T-cells

Virus

Viral antigen

Cell infected
by virus

Killer
T-cell

Infected cell is killed (lysed)
by killer T-cell

FIGURE 8. Cellular aspects (*left page*) and antibody aspects (*right page*) of immune responses.

Antibody immunity

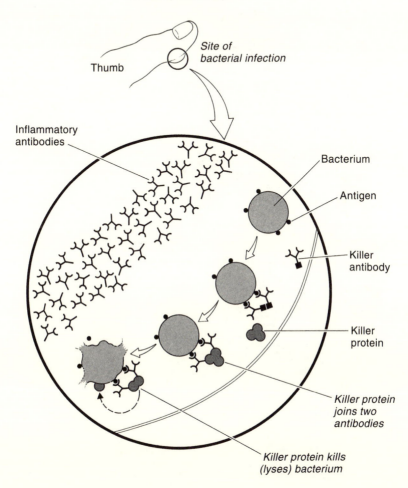

Thumb

Site of bacterial infection

Inflammatory antibodies

Bacterium

Antigen

Killer antibody

Killer protein

Killer protein joins two antibodies

Killer protein kills (lyses) bacterium

IgG activates inflammatory proteins and activates killer (lytic) proteins

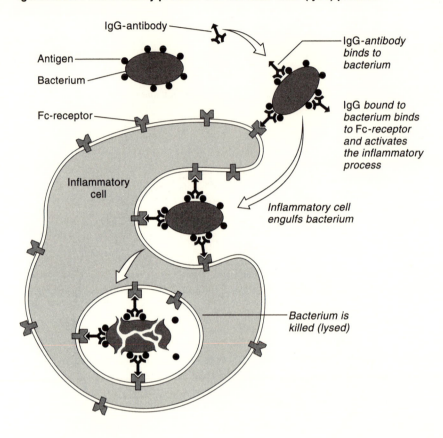

FIGURE 9. Different classes of antibodies (*this page and facing page*) have different functions.

IgE causes allergies and protects against parasites

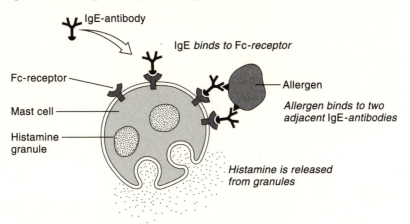

IgE-antibody

IgE *binds to* Fc-*receptor*

Fc-receptor

Allergen

*Allergen binds to two
adjacent* IgE-*antibodies*

Mast cell

Histamine
granule

*Histamine is released
from granules*

IgA protects secretory body cells against bacteria

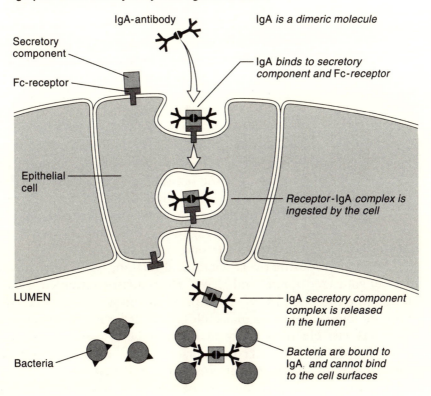

IgA-antibody

IgA *is a dimeric molecule*

Secretory
component

IgA *binds to secretory
component and* Fc-*receptor*

Fc-receptor

Epithelial
cell

*Receptor-*IgA *complex is
ingested by the cell*

LUMEN

IgA *secretory component
complex is released
in the lumen*

Bacteria

Bacteria are bound to
IgA. *and cannot bind
to the cell surfaces*

Kenichi Arai, accomplished in classical biochemistry and en-zymology, had acquired the techniques of molecular biology and was now becoming familiar with the assays, problems, and jargon of immunology and hematology. With his team of eager, talented young cloners, largely Japanese (Takashi Yokota and Atsushi Mi-yajima being the most prominent among them), he directed the power of recombinant-DNA technologies to isolate the mysterious growth factors and to push for the reduction of some of the key questions in biology and immunology to molecular terms. The fre-quent meetings Kenichi organized to mobilize and coordinate the teams from the cellular and molecular sides usually ended with him saying, "We can begin immediately," which was later used as a motto on a popular DNAX T-shirt.

The ambience at DNAX, a matchless blend of academic pursuit with industrial resources, was shaped and honed by people on both the science side and the industrial side, but especially by Allan Waitz. Installed by Schering-Plough as the CEO of DNAX, Waitz refused to impose the Schering-Plough company culture. Instead, he adopted the unregimented DNAX academic style and became its staunchest guardian and interpreter. In his administrative role, he interdigitated DNAX with the Schering-Plough system to make effective use of its resources while repeatedly hacking at the Schering-Plough tentacles that tried to envelope DNAX and assim-ilate it. DNAX was established at the outset of the relationship as a free-standing Schering-Plough subsidiary, and Waitz fought to keep it unencumbered by the regulations and patterns imposed on divisions of the company. On the occasion of a quarterly meeting of the policy board at DNAX, when the visiting Schering-Plough contingent caucused in advance with Waitz, one of the group was prompted to ask, "Are you working for us or for them?" Waitz took the inquiry as an unintended tribute to his success.

Without imposing overt scientific direction, Waitz appreciated and encouraged talent and drive. He nourished Kenichi Arai's leadership role and promoted the physical interactions between the immunologists and molecular biologists in order to create genuine team efforts. He read every manuscript before submission and made helpful and critical comments. His open-door policy set the tone for a trim and eager administrative staff that solved the per-sonal and practical problems of the scientists to free them to do their work.

In view of Waitz's imminent return to Schering-Plough in 1983, we tried hard, but without success, to recruit an immunologist of stature whose outlook would complement DNAX's strength in molecular biology, a scientist with the skills to lead and to expand the research enterprise. Unknown to us at DNAX, Schering-Plough management had offered the position to someone with scant scientific credentials, but, fortunately, this appointment was averted. After Waitz had returned to Schering-Plough in the middle of 1983, DNAX desperately wanted him back. Schering-Plough responded to our entreaties by persuading Waitz to become the permanent President and CEO of DNAX.

A Trough between Waves

After 1986, the wave of progress and the excitement of the early years began to subside. Some problems were evident, and others became apparent in retrospect. It would be at least two years before the trough of the wave would be succeeded by the next big crest.

There were a number of problems. Several cloned cytokines were in hand, but in very limited amounts. Relatively little was known about their receptors and their functions in cell culture, let alone in an intact mouse or human. Experience in physiological studies—scant at DNAX and virtually nonexistent at Schering-Plough—was difficult to acquire. Having inadequate supplies of the pure cytokines and not knowing how best to apply these novel and precious compounds to the treatment of disease was doubly frustrating.

Cloning and identification of new factors had become increasingly competitive. Biotech ventures, pharmaceutical companies, and academic institutes had acquired the new technologies and were adding innovative variations. Some ventures, like Immunex, targeted their considerable manpower and resources toward winning cloning races rather than pursuing basic studies to uncover novel factors and their functions. With more resources focused on defined targets, DNAX might have been the first to clone some of these cytokines, such as G-CSF, which is approaching one billion dollars in annual sales for Amgen.

Transfer of DNAX discoveries to Schering-Plough for development bogged down repeatedly for technical and organizational reasons. It was galling to see a competitor's group, apparently behind in the discovery of GM-CSF, move rapidly ahead in product development, clinical testing, and drug approval.

Within DNAX, dissension arose over discoveries made through team effort, particularly in the case of IL-4. Several groups had pursued one or another of its multiple functions, and each had contributed significantly. However, all were unhappy with their share of the credit. As a result of this experience, the scientists were inclined—and even encouraged—to seek new and separate directions. Besides, the consolidation of old findings always seems less exciting than the discovery of new ones. With DNAX's increasing strength in immunology and its maturity in molecular biology, Kenichi Arai's firm hand was needed less, and, to some, he seemed overbearing. His own research group, disproportionately large and almost all Japanese, formed a discrete domain within DNAX with less than optimal scientific and social interactions with others in the institute.

Postdoctoral fellows of high quality began arriving in increasing numbers, and they were being assigned research problems outside the sphere of their advisors' DNAX-oriented projects in order to insure their independence from the possible pressures of these projects. As a result, there was a proliferation of projects and a diffusion of effort and attention over a broad range of immunology and molecular biology.

These several developments at DNAX were likely the reflection of cycles encountered in all research programs—discovery, followed by further explorations, and succeeded by consolidation of the significance of the new findings. Based on the techniques, systems, and insights gained from that new knowledge, a program is poised for a fresh discovery. The crests and troughs are more prominent in a young institution than in a mature one, where two or more concurrent cycles are superimposed out of phase, giving the overall impression of steady, smooth progress.

In the period 1986–1989, at the young DNAX in the trough phase, the apparent lack of cohesion and the dispersion of efforts threatened the effectiveness of most projects and the success of DNAX as an entity. As a diffuse, miniature version of NIH, DNAX had no prospect of scientific distinction, nor could it expect to be retained as a permanent ward of Schering-Plough. These concerns were voiced by the scientists at off-site retreats, and a similar uneasiness was felt by some of us in our advisory roles. Altogether, there

Maureen Howard and Kevin Moore

was a lack of confidence that DNAX could repeat the triumphs of its early years.

During the summer of 1988, I interviewed each of the scientists to learn their views about the health and direction of DNAX, and to learn how they regarded their own research, status, and future careers. They gave remarkably similar assessments. Most or all of the aforementioned problems were recognized as contributing to the slowdown in productivity and the loss of enthusiasm. Not surprisingly, the scientists were unhappy with the increasing diffusion of research directions, and they eagerly proposed various remedies. At the same time, they were all grateful for the resources and the research opportunities at DNAX, and they expressed no wish to go elsewhere.

Then good things began to happen, building eventually to a new and larger wave of progress and confidence. On the administrative side, Maureen Howard, young and ambitious, arrived in 1986 from a postdoctoral fellowship with William Paul at NIH,

Jacques Banchereau (left), with Robert Coffman and Tim Mosmann

with expertise about B cells and their growth factors. She became the first director of immunology, bringing enthusiasm, cohesion, and visibility to the people and programs of the department. Reorganizations at Schering-Plough enabled Waitz to report directly to Alex Lane, President of Pharmaceutical Research, thus bypassing difficulties at intermediate layers of authority. T. L. Nagabhushan, given responsibility for biotechnology at Schering-Plough, was put in place to expedite the transfer and development of DNAX discoveries.

Advances on the scientific front were more significant. Mosmann, Coffman, Rennick, and Albert Zlotnik, tinkering with bioassays, came upon novel and potentially important growth factors. Internal alliances were arranged to clone these new factors in place of projects that were wide of the perimeter of DNAX interests. A landmark basic discovery reported by Mosmann and Coffman in 1986 bore fruit in 1989–1990, with the cloning, by Kevin Moore, of a powerful interleukin, named IL-10, an advance that galvanized all of DNAX as nothing had before.

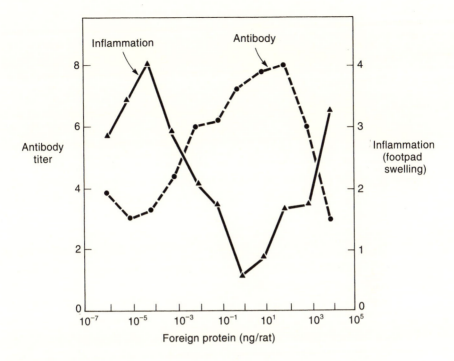

FIGURE 10. Distinctive and inverse immune responses to different amounts of a foreign protein (bacterial flagellin) injected into the foot-pads of rats (measured in nanograms per animal). Antibody titer is a measure of circulating antibody to the flagellin; the degree of footpad swelling is a measure of the inflammatory reaction. (From C. R. Parish and F. W. Liew, 1972. *Journal of Experimental Medicine*, vol. 135, p. 298.)

The IL-10 discovery contributed a key element to solving the puzzling observation of an inverse relationship between two distinctive immune reactions (Figure 10). Administration of a tiny amount of a foreign protein to a rat evoked a tissue (inflammatory) response, while very much larger amounts provoked a strikingly different humoral (antibody) response. Mosmann discovered that clones of helper T cells (as distinct from cytotoxic T cells, which destroy certain other cells) can be divided into two categories, TH1 and TH2, distinguishable by the cytokines they produce. Crucial to this discovery was Mosmann's knowledge of cell types, cytokines, and assays that had been perfected and made available by many scientists at DNAX.

The TH1 clones release IL-2, a tumor necrosis factor (TNF-β), and gamma interferon (IFN-γ), cytokines that lead to inflammation and the various activities that eventually destroy cells infected with viruses, bacteria, and parasites (Figure 11). The TH2 clones release IL-4, IL-5, and IL-13, cytokines that stimulate B cells to manufacture the antibodies that attack toxins, bacteria, and parasites *outside* host cells and that prevent new infections. The immune system generally makes an appropriate choice between the TH1 tissue-type and TH2 humoral-type responses. However, in some conditions—such as leprosy, infections with *Leishmania* (a flagellate protozoan parasite), and possibly AIDS—the selection of a TH2 response induces many features of allergy and may lead to more severe disease, and even to death.

Gamma interferon, the cytokine that promotes inflammation, was known also to suppress the antibody responses mediated by TH2 cells. Mosmann reasoned that there might be a reciprocal action, by a factor produced by the "antibody-helper" TH2 cells, that suppresses the inflammatory effects of the TH1 cells. When early evidence indicated the existence of such an activity, one possessed by no known cytokine, the factor was partially purified from mouse cells, and then, with Kevin Moore joining the project in 1989, the gene for the new factor was promptly cloned and the pure cytokine, IL-10, was isolated.

There were some anxious days for Moore and the rest of us when the protein sequence of IL-10, as inferred from its DNA sequence, revealed a most remarkable and, at that time, worrisome feature. Some 85 percent of IL-10 was virtually identical with a protein encoded by a well-known human virus, the Epstein-Barr virus (EBV), which is responsible for infectious mononucleosis and a tumor of the lymphatic system, Burkitt's lymphoma. "Please, not again, Lord," implored Moore. "Forget the ham and cheese, I just want out of this sandwich!" He had spent four fruitless years in collaboration with Teruko and Kimishige Ishizaka, notable immunologists at the Johns Hopkins University, tracking a factor that appeared to suppress the production of IgE, the allergy immunoglobulin responsible for allergic reactions. That factor turned out to be a protein component of intracisternal A particles (IAP), a collection of particles, resembling defective retroviruses, found

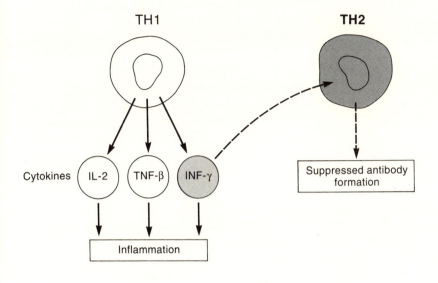

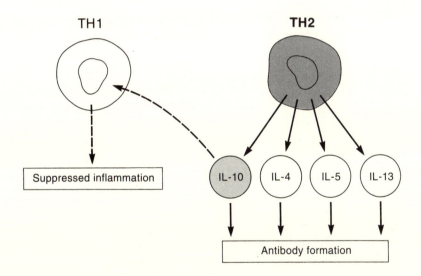

FIGURE 11. Two classes of helper T cells (TH1 and TH2) produce
distinctive cytokines and cross-regulate each other: IFN-γ produced by
the TH1 cells suppresses the functions of TH2 cells, whereas IL-10
produced by the TH2 cells suppresses the functions of the TH1 cells.
(TNF-β is a tumor necrosis factor.)

only in mice and rats and with no likely physiological significance in the regulation of IgE and allergy.

Several aspects of mouse IL-10, despite the presence of an EBV-related component, were reassuring. Since EBV infects only primates, the mouse seemed unlikely to have encountered this virus and to have been marked by it. Furthermore, the functions of IL-10 fit nicely into a scheme, anticipated and then confirmed, for its role in balancing immune responses to infection. Finally, when Moore succeeded, later that year, in cloning and isolating IL-10 from human T cells uninfected by EBV, he found it to be nearly identical with the mouse version. Now it could be understood that, in the evolution of EBV in primate hosts, a major part of the IL-10 gene had been captured by the virus and used in a manner that promoted its transition from a pathogen to a benign parasite.

The IL-10 discovery excited and immediately involved virtually all the scientists at DNAX. The immunologists, working on mouse T cells, B cells, bone-marrow cells, and thymus cells, explored the functions of IL-10 alone and in combinations with other cytokines under a variety of conditions. Monoclonal antibodies were prepared for analysis and the neutralization of IL-10. Studies were started to determine how IL-10 inhibits the release of cytokines from TH1 cells. The immunologists focusing on human cells were highly skeptical that the helper-cell dichotomy in mice would apply to humans, but, with sustained effort, they and others finally proved that it did. The molecular biologists sought improved vectors and expression systems to produce IL-10 in greater quantity, examined the structure of the cytokine in finer detail, and set about to alter the structure genetically to determine which features are responsible for each of its several functions—and much more.

The team efforts on IL-10 produced a torrent of publications in the most prestigious journals. In contrast to the dissension over the credit for the IL-4 discovery, there was a widely shared sense of achievement and pride that DNAX had not only succeeded in science but had provided Schering-Plough with a product of great clinical potential. Schering-Plough management, initially bored with "still another interleukin," was soon suitably impressed and quickly deployed teams from all its divisions. Nagabhushan directed the improved production of IL-10 for further biological studies at DNAX, arranged for crystallographic examinations by Charles Bugg at the University of Alabama, and helped organize efforts in product development, clinical testing, regulatory approval, and marketing.

As with other cytokines, having the pure compound and find-ing the gene that directs its synthesis opened the way to answering three major questions: How does IL-10 affect a particular cell or tissue? How does IL-10 affect the whole animal? Of what value is IL-10 in the treatment of disease?

When IL-10 was added to the culture medium in which each of a variety of cell types was grown, it became clear that, in addition to promoting the production of antibodies by B cells and suppress-ing the release of inflammatory cytokines by TH1 cells, the cyto-kine also enhances the growth of many kinds of blood-forming cells. In probing how IL-10 exerts its action on a cell, Kevin Moore identified the receptor molecule on the surface that binds to the cytokine and sends a signal to the DNA in the nucleus to activate certain genes and to replicate the DNA (Figure 12). Knowing the

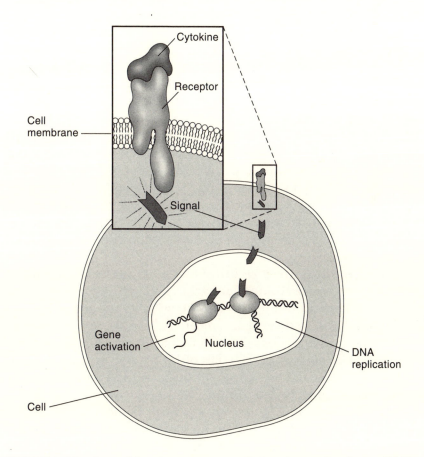

FIGURE 12. A cytokine, its receptor, and signal transduction to DNA.

IL-10 receptor, he could determine which cells in the body possess it and provide Schering-Plough with an assay to discover drugs that can substitute for IL-10 or antagonize its action by competing for the receptor. The nature of internal signals and how they are transmitted—called signal transduction—are issues at the forefront of current research at DNAX and worldwide.

How a cytokine affects the whole animal can best be assessed by removing it, either in the adult animal, by administering a neutralizing antibody, or in embryonic life, by knocking out the gene that encodes it. In the case of mouse IL-10, an antibody against it (prepared in rats and then administered to mice) showed some of the anti-inflammatory effects anticipated from cell-culture studies. Surprisingly, a certain class of B cells (Ly-1 cells, which are responsible for a number of distinctive immune responses) was absent, apparently because IFN-γ, which can suppress these B cells, was not held in check by IL-10.

A more striking result of removing IL-10 came from mice whose IL-10 genes had been knocked out. They were scrawny and anemic, and they died within a few months of severe inflammatory bowel disease. While the importance of IL-10 as a regulator of the inflammatory response had been anticipated, the particular localization of severe reactions—tissue breakdown and hemorrhage in the intestinal tract—had not. Clearly, the elaborate immunity apparatus of the gut—greatly underappreciated when compared with other lymphoid systems in the body—must be a major bastion against invasion by a vast array of bacteria, fungi, and other pathogens. In fact, when mice deficient in IL-10 were raised in a germ-free environment, their bowel disease was limited and less severe. Clearly, IL-10 is an essential regulator of immune responses in the intestinal tract.

The indications from cell-culture studies and animal studies have suggested several clinical uses for IL-10. Of immediate interest is the treatment of inflammatory bowel diseases, of which the debilitating Crohn's disease and ulcerative colitis are the most serious. Other inflammatory diseases for which effective treatments are inadequate and for which trials of IL-10 are contemplated are rheumatoid arthritis, psoriasis, and myositis. Based on the properties of IL-10 observed in cell cultures, uses may also be found in maintaining tissue transplants, in warding off the toxic effects of grafted tissue, and in treating autoimmune diseases and other disorders of the immune system. The low toxicity of high doses of IL-

10 observed in preliminary (phase I) clinical trials encourages efforts to explore its value in the treatment of a variety of serious diseases.

The discovery of IL-10, at the time it was made, could have come only from DNAX. The prompt coalescence of all the scientific resources of the Institute to exploit the discovery would have been possible nowhere else. Such sharing of knowledge, resources, and effort would never happen in a university setting, where each research group is discrete, self-contained, and jealous of its own status and security. Nor did any pharmaceutical company have the concentration of talents or the organization to make the IL-10 discovery and then to expand it rapidly and effectively. This was DNAX at its best.

As much as any other development at DNAX, the IL-10 discovery proved that the sustained pursuit of a basic question in science, supported by excellent resources in a cooperative, collegial atmosphere, can lead to a product of great commercial potential. The success of this effort validated Schering-Plough's unwavering and unstinting commitment to the long-term support of basic research at DNAX. For the rapid expansion and development of the IL-10 discovery, credit belongs to all the DNAX scientists, for their immediate and intense involvement, and to the Schering-Plough managers and staff, who responded promptly and appropriately at every level.

The DNAX Route to Academia

Many young scientists who launched their research careers at the NIH achieved enough, in just a few years, to be offered professorial posts in outstanding university departments. This "NIH shunt" around the academic ladder has been repeated at DNAX. The academic focus of DNAX was validated by the U.S. Department of State in its approval of an Exchange Visitor Program with the granting of J-1 visas to foreign guests, a status virtually unique among industrial institutions in the San Francisco Bay Area.

Although scientists recruited to DNAX could not be offered a conventional tenure-track position, they were awarded the more important security that is earned by productivity and visibility in science. At DNAX, we sought to ensure that the energy and creativity of the scientists be the determining elements of their future.

The ambience at DNAX can be described as an opportunity to work at a particular forefront of science, with excellent resources, in a free and open atmosphere, and with the support of, rather than intrusion by, administrative staff. An important constraint is that the focus of research be in productive areas of immunology and molecular biology. Another is that the scientist not be reclusive and have concern for the success of the communal enterprise. The group size of six to ten people for a senior scientist may appear small when compared with larger units in other settings, but this limitation is largely offset by the wide range of resources made available for biological, analytical, and preparative studies.

How open is DNAX? Visitors enter unheralded and unbadged, and they have free access to the laboratories. There are no pre-employment drug screens, dress codes, or strict notebook protocols. With the labs active days, nights, and weekends, with equipment operated in crowded hallways, with daily seminars and discussion groups, and with a constant torrent of publications, the atmosphere reflects a work ethic driven by ambitious individuals rather than by an organization, academic or industrial.

In my view, DNAX is an academy, a company of scholars searching for new knowledge with minimal constraints. I recall an experience in 1947, when I left the embrace of the laboratory of Carl and Gerty Cori at Washington University in St. Louis to return to the NIH. Gerty despaired that my career in science would be blighted by my (willing enough) consignment to a government laboratory. When she visited me a year or so later, I tried to convince her that, with the generous resources available to me, I could deepen my new-found love of enzymes in the stimulating company of like-minded colleagues—Leon A. Heppel, Bernard L. Horecker, Herbert Tabor, and others. Working full tilt, and with the four of us conducting *daily* lunchtime seminars, I was enjoying an ideal academic atmosphere. I would later look back wistfully on those NIH years between 1947 and 1952, before taking on the chairmanship of the Microbiology Department at Washington University, as truly blissful.

On the occasion of the tenth anniversary of the DNAX laboratory in 1991, we could feel pleased that the scientists at DNAX were enjoying a genuine academic atmosphere, because they work, with optimal resources, on the problems that are of keenest interest to them. They do so in the company of colleagues of comparable ability and outlook, and they derive satisfaction from contributing to the success of a communal effort. With regard to the application of their laboratory discoveries, they can, in good conscience, leave them in the hands of others at Schering-Plough who are competent to develop them into products of medical and industrial value.

At DNAX, scientists are emboldened to pursue high-risk problems for a long period, assured that they will be judged by the quality of their effort rather than by publications and approbation (by peer review) of grant applications. Kevin Moore, who spent seven years on three different problems with disappointing results before making spectacular progress in the cloning and isolation of IL-10 and its receptors, mouse and human, is one (but not the only) good example of this DNAX style. Freedom from dependence on grants saves the inordinate amount of time, and the anxieties, that such applications require, as well as the need to justify relevance to a targeted disease.

DNAX also provides generous fellowships for two to three years for some sixty outstanding young people to work under the direction of twenty-five scientists. The responsibility for their guidance is rewarded by the stimulation and enthusiastic effort of the fellows, who bring a fresh variety of skills and attitudes and the drive to do research notable enough to earn staff positions elsewhere. Job placements have been about equally divided between university departments and industry.

Unencumbered by teaching loads, committees, and bureaucratic obstructions, given access to all levels of a lean and responsive administrative staff, afforded excellent salaries, benefits, and bonuses, DNAX scientists are largely freed from immediate financial concerns and the diversion of seeking or weighing offers from other institutions. With all that, a professorial status, with its opportunities to teach in basic and advanced curricula, collegial interactions with other academic disciplines, and roles in maintaining traditions of scholarship, can be a strong lure for some. Although the path from academia to industry has historically been irreversible, exceptions have cropped up, most notably at DNAX. In view of the strong academic atmosphere at DNAX, the transition to a university setting seemed less sharp to four DNAX scientists—Kenichi Arai, Takashi Yokota, Kunihiro Matsumoto, and Tim Mosmann—who decided to return to academia.

"How old are you, Kenichi?" I asked in 1986. "In three years, I will be 46," was his response. He regarded 46 as the "point of no return," the age when scientific, institutional, and personal considerations would militate against his returning to Japan to assume a professorship at the University of Tokyo, his alma mater. He had left the Institute of Medical Sciences of the University of Tokyo (IMSUT) for DNAX in 1981, because he wanted to switch from classical biochemistry in bacteria to studies of DNA replication in eukaryotic organisms, combining the newer disciplines of molecular and cell biology.

Having been impressed by the democratic organization of the Biochemistry Department at Stanford during his earlier fellowship stay, Kenichi Arai wished to create and enjoy a similar atmosphere at IMSUT when he returned in 1981. But these ambitions were impossible to achieve in Japan with the very limited resources afforded an assistant professor and the rigid hierarchical structure of a Japanese department. Given the possibility of attaining these objectives at DNAX, he was willing to endure the stigma attached to joining an industrially sponsored enterprise. It was an enormous gamble, but his talent, dynamism, and leadership helped make it succeed.

Arai saw that Schering-Plough's proposal that DNAX focus on T cells created an opportunity to explore how their activation by an antigen or another signal was transferred from the cell membrane to the nucleus to initiate DNA replication and cell division. This goal would ultimately remain beyond his reach, but his enthusiasm impressed the Schering-Plough teams visiting DNAX in 1982 and was instrumental in their recommending its acquisition. By his aggressive application of expression cloning to the isolation of growth factors identified by the immunologists, novel cytokines were discovered, and their physical and functional characteristics were defined. With this, Kenichi Arai's scientific reputation grew as a pioneer in molecular immunology, and his exhaustive, comprehensive seminars demonstrated a remarkable breadth of view as well as detailed knowledge of the subject.

Arai was just as active in the governance of DNAX, organizing research teams, moderating disputes, reviving spirits, and firing off lengthy memos in all directions. Most memorable was a salvo in 1985 to Luciano, the Schering-Plough CEO, critical of jealousies and bureaucratic obstructions at Schering-Plough, citing the importance of keeping DNAX science separate from product development, and warning that increasing competition in industry and academia demanded more support for DNAX to maintain its competitive edge. In response, Luciano reaffirmed, as he had to others of us, his full support of DNAX.

The path to academia was wide open to Kenichi Arai, particularly to professorial posts in Japan, where respect for DNAX had replaced the earlier skepticism and Arai's share in its reputation was widely recognized. Finally, patriotism and loyalties led him, in 1989, to accept a professorship in the Department of Molecular Biology and Developmental Biology at IMSUT and a role in directing a series of appointments at the university that were designed to restore Tokyo to its once premier status in biological and medical science. Returning with him was Takashi Yokota, who had come to DNAX as a postdoctoral fellow, and now, hardly eight years later, was assuming the sole associate professorship in Kenichi's department, a post that would normally require an apprenticeship twice that long.

How ironic that, in returning to the University of Tokyo to accept a professorship at IMSUT, Kenichi would exchange places with Kaziro, his mentor, who had so disfavored Kenichi's move to DNAX. Kaziro, upon his retirement from IMSUT and the university, found the offer of a position at DNAX, with the title of Distinguished Visiting Scientist, to be the most attractive for continuing his academic life. In 1993, after Kaziro had been three years at DNAX, Schering-Plough enabled Kaziro to resume his research in Japan by establishing a special unit, called a Donation Department, for him at the Tokyo Institute of Technology, providing five years of full support for him and his group.

Although Naoko Arai had received an offer of a professorship at Yokohama University, her alma mater, a most unusual distinction for a woman in Japan, she still chose to remain at DNAX and to continue the exploration of how cytokine genes are turned on and off. With their Americanized son, Kensuke, in high school, daily fax exchanges and frequent visits from Kenichi keep the family intact.

Kunihiro Matsumoto held the lowly title of instructor in the Department of Engineering at Totori University, a small provincial institution in Japan. His work in yeast genetics caught the attention of Michael Wigler at the Cold Spring Harbor Laboratory, who enlisted James Watson, the director of the laboratory, to recruit Matsumoto as a postdoctoral fellow. Arai asked me to intervene to divert Matsumoto to DNAX. I was able to convince Professor Yasuji

Oshima, his mentor at Osaka University, of the greater resources and independence Matsumoto would enjoy at DNAX, compared with the close direction and relative anonymity of Wigler's retinue of quiet, industrious Japanese postdocs at Cold Spring Harbor, so vividly described in Natalie Angier's book, *Natural Obsessions*.

At DNAX, Matsumoto's work came quickly into full bloom. Afforded the necessary time, facilities, technical support, and stimulation, he demonstrated how manipulation of yeast genes can provide insights into growth and development (and oncogenetic aberrations) in mammalian cells. Although quite isolated from the immunologists by his retiring personality and language difficulties, he managed to interact with the molecular biologists enough to extract ideas and reagents and to impart to them the power of genetics and the advantages of working with yeast. Although his direct contribution to DNAX projects was minimal, his raft of important papers and his very presence gave DNAX wide recognition in the circles of cell and molecular biology. Within only five years, Matsumoto was catapulted to a full professorship in genetics at Nagoya, one of the elite Japanese universities, an advance that would have been virtually impossible under any other circumstance.

Tim Mosmann agonized for more than a year before deciding to leave DNAX and assume the chairmanship of the Immunology Department at the topflight University of Edmonton. He had come from there eight years earlier in the first wave of DNAX scientists. Imaginative, stimulating, and resourceful, he made key contributions to DNAX's success. At the outset, he alone was familiar with T-cell cultures and was able to improve bioassays for the rapid screening and characterization of growth factors. He invented a color-based assay for cell proliferation that largely replaced the time-consuming method based on the incorporation of radioactive thymidine, an assay that became widely adopted. His discovery of the dichotomy of helper T cells is surely one of the major advances in immunology of the past decade, his review in 1989 being the fifth most cited paper in all of biological research for that year. From these basic studies emerged the discovery of IL-10, which is likely to be one of the most lustrous gems among the cytokines in its clinical impact.

Mosmann needed DNAX as much as DNAX needed him. He could not have made those singular discoveries at Edmonton or anywhere else at the time. The collection of cultures, antibodies, cytokines, colleagues, and, most significantly, the climate of sharing enabled him to do this monumental work with a group of only four or five associates. At Edmonton, with resources to accommodate a research group several times that size and with students to teach, he must also be the custodian of a large department and be involved in the medical school and university budgets, curricula, and politics. For the development of discoveries, he must write the business plans to attract potential investors. Which is the true academy, DNAX or the university department?

Among the candidates to fill the openings created by the departures of Arai, Yokota, Matsumoto, and Mosmann, some were tenured at universities and others, who were just starting out, preferred DNAX over academic alternatives. Those chosen brought new and complementary interests to DNAX and an eagerness to adapt to its style. Four outstanding young scientists, with competitive offers of attractive academic posts, were recruited in 1993: Anne O'Garra, in the area of T-cell recognition; Tom Schall, to investigate chemokines, signal transduction, and cell trafficking; Emma Lees, to explore cell-cycle control; and Fernando Bazan, to be a resource for structural biology.

The senior scientists who remained at DNAX in 1994 continued to receive invitations to professorial posts and attractive offers from competing biotech companies. Atsushi Miyajima, torn between staying on at DNAX to extend his world-class work on cytokine receptors and returning to Japan, decided for personal reasons to take the offer of a professorship at the University of Tokyo. Frank Lee, who had emerged from Kenichi Arai's shadow to direct the molecular biology division, accepted the challenge of becoming research director and vice president of Millennium, a new biotech venture in Cambridge, Massachusetts. Remaining at DNAX, Donna Rennick has become expert in the biology of hematopoiesis, and Bob Coffman has extended his experience in molecular biology to even more complex parasitic and other diseases. Gerard Zurawski has assumed the direction of a new DNAX division that will include studies of the structure–function relationships of the cytokines, to which he has contributed greatly.

A sampling of comments heard from them and others: Miyajima says, "I would not have been able to undertake such a speculative and risky project in any academic lab." Moore says, "I feel I have far more freedom at DNAX than I would ever have in academia." And from Rennick, "Each morning when I come to work, I say, 'Thank God for DNAX.'"

CHAPTER 6

The DNAX–Schering-Plough Connection

I have searched for an analogy in human relationships to frame the connection between DNAX and Schering-Plough—parental, fraternal, avuncular, friendly—because so much of its success rests on mutual trust and affection. But the choice of any analogy would likely distort more than it would clarify. What needs to be recognized about the connection is that it depended on a web of underlying personal interactions, including those particular to each circumstance, which mattered as much as principles and procedures and sometimes a good deal more.

The DNAX–Schering-Plough Connection: Past, Present and Future

Starting with the four founders of DNAX—Paul Berg, Charles Yanofsky, Alex Zaffaroni, and me—trust and affection shaped a unity of purpose and style. Within the hierarchical Schering-Plough management, there also prevailed a sense of mutual confidence and amity that was forged from sharing in the transformation of the company. The connection between DNAX and Schering-Plough, from its outset, was built on shared respect between the top Schering-Plough executives—Bob Luciano and Hugh D'Andrade—and Alex Zaffaroni. The future success of an expanding and complex organization will require that these combinations of personal and professional respect and affection be extended to the DNAX and Schering-Plough people who had no role in building the early relationships.

159

The DNAX style, fashioned from the devotion of the Stanford founders to sustained basic research, was readily accepted and implemented by Zaffaroni. The theme of his many ventures, from Syntex to Affymax, has been—and still is—technologic innovation for applications to marketable medicinal products. He understood the need of scientists to follow their curiosity and could, as he had in his own career, orient their activities toward industrial goals. But the DNAX venture differs fundamentally from his others. The fiscal security derived from the Schering-Plough connection provided DNAX the freedom to pursue basic knowledge at a forefront of science with broad, long-term objectives.

On the Schering-Plough side, the dominant influence was the vision and courage of Bob Luciano. His grasp of the importance of biotechnology for the company's future and his willingness to gamble on a distant payoff were basic to his decision to invest in DNAX. Beyond that, he managed to enlist his able and loyal managers in shaping this unusual academic-industrial liaison. Hugh D'Andrade, who was responsible for strategic planning and the crafting of the DNAX acquisition, became chairman of the joint policy board. Richard J. Kogan (the chief operating officer), Alexander Z. Lane (in charge of all pharmaceutical research), and Frank Bullock (directing the discovery division) handled Allan Waitz's proposals for the budget, the capital expansions, and the head count at DNAX.

Key to the success of DNAX was expanding the good rapport between Zaffaroni and Luciano to embrace the managers and scientists at Schering-Plough and DNAX. Frankness, understanding, and respect in discussions at board meetings were supplemented by telephone exchanges over unresolved or worrisome issues. Convivial dinner parties in New Jersey, New York, and Palo Alto, with spouses included, helped to mature business relationships into firm friendships. Invitations to the DNAX founders and directors to join the Schering-Plough board of directors at meetings in elegant settings in France and Ireland helped bond DNAX to the board. It was also a lifestyle with which we did not mind becoming acquainted.

Education was ongoing with both sides. Those of us at DNAX, with little knowledge of the pharmaceutical business, learned about the complexities of large-scale production, the extraordinary quality standards for drug products, the hurdles of regulatory approval, and the uncertainties of marketing. With this experience came respect for the risks that had been taken and for the skills and

James Florio (left), with Richard Kogan, Hugh D'Andrade, and Robert P. Luciano

New Jersey Governor James J. Florio (right) and Robert P. Luciano viewing exhibits in the lobby of Schering-Plough's Drug Discovery Facility in Kenilworth, N.J., in 1993

persistence of those in the company who managed this complex operation and excelled in their essential jobs. Schering-Plough had correspondingly little awareness of the work and habits of creative scientists, the importance of an open, nonsecretive atmosphere, the cost-effectiveness of innovative basic research, and the rapidly expanding frontiers of molecular biology. The managers, both lay and scientist, were quick learners of the novel science; they assimilated the DNAX discoveries, and they became supportive of its style.

When Schering-Plough acquired a new laboratory facility for DNAX, eventually enlarged to more than 70,000 square feet, proximity to Stanford was the most significant factor. The Schering managers acceded to our wish to locate the DNAX laboratories in the Stanford Industrial Park adjacent to the campus, even though the cost was twice that of a comparable facility 20 minutes away, either in Redwood City to the north or in Mountain View to the south. (By contrast, the half dozen Schering-Plough facilities for research, production, and administration in New Jersey are separated by 30-minute drives on busy highways.)

Having Stanford scientists and seminars close by helped in the recruiting of university-oriented young scientists. It also meant that Berg, Yanofsky, and I could attend DNAX meetings and seminars as easily as any engagements on the campus. The DNAX relationship was also of direct benefit to Stanford. With the growth of science and resources at DNAX, key reagents and procedures developed there were sought by Stanford scientists; increasingly, DNAX gave more than it received.

What has Schering-Plough done for DNAX, and what has DNAX contributed to Schering-Plough? At DNAX, the strong and sustained Schering-Plough support has advanced both science and the careers of young scientists. DNAX now enjoys world-class stature in molecular immunology, and it attracts the finest newly minted Ph.D.'s for postdoctoral training. Some 200 of them have come through DNAX to take places eventually in research, roughly half in academia and half in industry.

Virtually all DNAX scientists have accomplished enough within a very few years to be offered attractive posts in academic or industrially sponsored institutions. The setting provided by DNAX has enabled these ambitious young people to join in communal projects and to feel rewarded by DNAX's success in science and in its transfer of knowledge to promote human welfare. From 1986 to 1990, DNAX was the world's most cited institution—government, industrial, or academic—in publications in the field of immunology. The high quality of science and training and the ambience created at DNAX have been a source of pride for its founders and a model of research organization for academia and industry to emulate.

The contributions of DNAX to Schering-Plough have been both tangible and intangible. Among them:

- GM-CSF is now marketed (as Leucomax) in Europe and Israel with annual sales of $44 million.

- IL-4 and IL-10, with potential for treatment of major diseases, are now in early clinical trials.

- The DNAX pipeline has novel cytokine candidates and receptors that may provide assays for the discovery of orally active, small-molecule substitutes for the cytokines and for the molecules that inhibit their actions.

- Leadership in pharmaceutical biotechnology derived from DNAX has enhanced Schering-Plough's image in the investment community, has helped it to recruit high-quality scientists and managers, and has attracted initiatives for joint ventures from top-ranked scientists.

- Interactions with DNAX have elevated the quality of all Schering-Plough research programs by accelerating the introduction of advanced techniques of molecular biology. As an example, screening for receptor substitutes and inhibitors has been highly refined by substituting pure receptor molecules in place of the crude tissue preparations used in classic pharmacological assays. Spurred on by the DNAX example, Schering-Plough scientists have developed greater stature in the scientific community and greater influence with their managerial hierarchy.

- The Unicet (Schering-Plough) laboratory in Dardilly, France, previously in disarray, was reshaped into a highly productive human-immunology research institute as a DNAX subsidiary.

- Business opportunities in Japan and elsewhere have been opened by DNAX's reputation and by introductions made by its alumni abroad.

- The dollar value of DNAX, were it sold to another company, would likely far exceed the Schering-Plough investments.

In striking contrast to the DNAX connection stood Schering-Plough's previous attempt to support an outlying basic-research operation, which was destroyed by poisonous personal frictions. No one in Schering-Plough's current management knows much about RIMAC (Research Institute of Medicine and Chemistry), which was founded in 1958 in Cambridge, Massachusetts, by Maurice M. Pechet with the support of Schering Corporation and the blessing of Francis Brown, then its CEO. Four years earlier, Pechet, a steroid chemist trained by Louis Fieser, and later a Harvard M.D., had demonstrated, while at the National Institute of Arthritis and Metabolic Diseases, the greater clinical effectiveness of the new Schering metacorticosteroids, prednisone and prednisolone. Gratitude for the profitability of these drugs and an intimate rapport with Pechet, rather than any long-range view of the benefit of academic-style research, impelled Brown to finance RIMAC. The annual budget of nearly $500,000 (a handsome sum in 1958) from Schering sustained the laboratory, which was located in an old cough-mixture factory and had a staff of about fifteen under the guidance of Sir Derek H. R. Barton, a Schering consultant in steroid chemistry.

Synthesis of aldosterone, the previously inaccessible adrenal hormone, was achieved at RIMAC by a novel Barton technique, but, unfortunately, it proved to be without significant therapeutic value. The RIMAC connection began to unravel in 1966, after the accidental death of Brown and the accession of W. H. Conzen to the helm of Schering. Hostility between Pechet and the Schering scientists had grown over the years from a mutual lack of scientific regard and Brown's deference to Pechet, to whom he had repeatedly offered the directorship of research at Schering. The deep resentment of Schering's scientists, kept out of the inner loop, and the lack of any drug leads from RIMAC, led to the gradual phasing out of Schering's support by 1973. The enterprise was doomed by lack of strategic planning and the absence of sustained personal trust.

Schering-Plough Middle Management

As in all marriages, there were sticking points, some minor and others of real consequence. At the outset, Schering-Plough insisted on milestones of progress at DNAX, the attainment of which would be required for an annual payout of Schering-Plough stock. I pleaded, but to no avail: "If research were that predictable, it wouldn't be basic. Serendipitous discoveries play a major role in science. Milestones are a charade." In each of these early years, we went through the ritual of setting out the milestones and then regularly judged that those of the previous year had been met, even though the actual achievements were quite different and more important.

Some problems with the management at Schering-Plough were more serious. After more than 15 years of jumping through the hoops of its bureaucracy, Al Waitz would not continue, on principle, doing that at DNAX, asserting that it should be treated as a subsidiary rather than a division of the company. In New Jersey, resentment flared over what appeared to some to be the pampered status of DNAX scientists and staff—freed of all commercial responsibilities, consuming gobs of money, and following their own fancy in a laid-back California style. Waitz, with all his experience, could navigate the bureaucratic labyrinths and, at key checkpoints, threaten an obstructive gatekeeper with a direct appeal to Luciano. But political agility alone succeeds only to a point. He could, and did, endure repeated frustration and bad annual performance reports, but gross failures at Schering-Plough to implement DNAX discoveries threatened DNAX's survival.

DNAX's discoveries of GM-CSF and IL-4 had languished, although they had been transferred promptly to Schering-Plough for development. Competitive biotech companies were gaining ground. DNAX scientists, dispirited by seeing their contributions neglected and mishandled, were also aware that, in the ultimate reckoning, the lack of profitable products would be attributed to them and not to those who had dropped the ball.

Many reasons could be given for Schering-Plough's failure to develop DNAX discoveries aggressively. The New Jersey research staff was busy with its own projects and the development of licensed products. They faced deadlines for submission of the number of INDs (investigational new drugs) and NDAs (new drug applications) that had been assigned them as milestones by upper management. Poorly versed in immunology and molecular biology, the Schering-Plough scientists could be excused for believing that their own inventions were more deserving of attention than those emanating from DNAX. Nor was there any pressure from the clinical or marketing divisions, where the focus of the established disease teams was on allergy, inflammation, infectious diseases, and cardiovascular diseases, and where there was no knowledge of, and no interest in, the strange DNAX proteins that promoted the growth of mouse cells in culture. DNAX would have taken on the exploration of cytokine functions in animals, but it lacked the pilot-plant facilities to produce the large amounts of protein needed for such studies. No help was forthcoming from Schering-Plough, because the production facilities there had been pre-empted by alpha interferon (Intron A) and other marketed products in short supply.

The managers at Schering-Plough responsible for DNAX operations were Alexander A. Lane and Frank J. Bullock. Lane, from 1980 on, was President of the Pharmaceutical Research Division (renamed, in 1992, the Schering-Plough Research Institute), responsible for the discovery, development, clinical trials, and, ultimately, the regulatory approval of all of the company's prescription pharmaceuticals.

When Alex Zaffaroni first pitched the DNAX proposal, Schering-Plough management decided that, if DNAX were to be acquired, its operation and budget would come under Lane's authority. In this context, Lane was a central member of the team that visited and evaluated DNAX. His favorable impressions of DNAX's people and research style were influential in the Schering-Plough decision.

Alexander A. Lane

From the mom-and-pop drugstore of Alexander Lane's parents to his marriage to Linda, a pharmacist in a family of pharmacists, and on through his schooling and jobs, medicinal drugs were at the center of his world. The oldest of eight children in a family of Polish origins, Lane helped out in the store while attending Jesuit schools in Detroit from the grade-school level through college at the University of Detroit. His classical education, with its minimal emphasis on science, was leavened by a curiosity for the magic in all those drugstore bottles and the romance in the lives of the scientists described by Paul de Kruif in *Microbe Hunters*.

Scholarly and determined, Lane went on to do graduate work in biochemistry in the Edsel Ford Research Institute at Wayne State University. He studied enzymes, particularly those that

Alexander A. Lane

operated in various cellular compartments in response to the metabolic needs of the body. With a Ph.D. degree, he realized how useful medical training would be for doing research on humans, and, especially, the advantage gained by having the M.D. credential, which would allow him to direct medical research. As a student at Wayne State Medical School and in the general internship that followed, he was preparing for a career in clinical research rather than practice.

With his roots in Detroit and his wife employed at Parke-Davis (as were her father and uncle), it was natural for Lane to take a job with the same company, first as a clinical investigator for two years and then, for four more years, as Director of Clinical Pharmacology. Some of those years were spent working at the state prison in Jackson, where he conducted clinical and laboratory analyses of early clinical trials of Parke-Davis drugs. His rapport with the prison subjects was good enough that they would warn him in advance of a planned riot.

Until 1966, at age 37, his environment had been limited to pharmacists, drugs, and Detroit. At the urging of Bert Peltier, a friend who had gone from Upjohn to Bristol-Myers in Schenectady, New York, Lane decided to move to an administrative position there, advancing, over the next eleven years, to Vice President and Medical Director. Reporting to Amel Menotti, his responsibilities included regulatory affairs as well as the acquisition of licenses for the development of promising new drugs. Herman Sokol, the CEO of Bristol-Myers, had extended the company's established focus on antibiotics to drugs for cancer chemotherapy, partly as a result of his own struggle with cancer. Among those anticancer drugs were bleomycin (obtained from Professor H. Umezawa at the University of Tokyo) and cisplatin (obtained from Dr. Barnett Rosenberg at Michigan State University).

With Irvan Pachter succeeding Menotti, who had retired in 1974, there was a progressive emphasis on marketing over research and science, a climate that Lane found increasingly inclement. The speed with which he had obtained FDA clearance for amikacin, a licensed Japanese aminoglycoside antibiotic, brought him to the attention of W. H. Conzen, who offered him a position at Schering-Plough. He began work at Schering-Plough in 1977, first as Senior Vice President of Research Operations, and, three years later, he replaced Douglas Lawrason as President of the Pharmaceutical Research Division.

When Lane arrived, the atmosphere of the research organization at Schering-Plough was desolate. There was little communication among the science departments in the division and even less with other divisions of the company. Motivation, morale, and competence were generally at a low level. For help in rebuilding research and development, Lane brought in Frank Bullock from Abbott Laboratories. Bullock had had extensive earlier experience at Arthur D. Little and in academic departments.

Frank Bullock, from 1981 to 1993, was Senior Vice President of Research Operations for Schering-Plough. Under Alex Lane's authority, he was directly in charge of DNAX. In the broadest sense, it was his job to close the gap between the chemistry and biology cultures within the company.

Bullock's extensive training and experience in a wide range of pharmaceutical disciplines, along with his managerial qualifications, had brought him an invitation from Lane in 1979 to join Schering-Plough as Vice President of New Drug Discovery. His charge was to upgrade the quality of Schering-Plough research and to formulate strategic research plans that could be articulated to the investment community by the CEO. It was in this role that Frank was called upon to assemble the task force in immunology that would set the stage for Schering-Plough's acquisition of DNAX.

Frank Bullock

Like Alex Lane, Frank Bullock had grown up in his father's drugstore business. He attended the Massachusetts College of Pharmacy in Boston, near his birthplace in Brookline and his home in Dedham. With no inclination to follow in his father's footsteps, but with a persistent interest in medicinal chemistry and research, he thought of going to one of the nearby medical schools, but he lacked the liberal arts courses to meet the admission requirements. Instead, he applied to the Chemistry Department at Harvard University and was astonished to be accepted. He became the last of the several hundred graduate

Frank Bullock (right) *with Charles Yanofsky*

students of the renowned Louis Fieser of steroid and textbook fame.

Fieser's research was the most drug-related of any member of the faculty. In his consulting role for Merck in the therapy of coccidial (protozoan parasite) infections, quinones were the prime focus, and their synthetic chemistry was the basis of Frank's thesis. He could have taken a job with Merck on the strength of Fieser's ties with Max Tishler, who was Merck's director of research. But Bullock was unimpressed by the empirical nature of drug discovery and more interested in a rational and physical understanding of molecular interactions.

An NIH postdoctoral fellowship in 1964 with Oleg Jardetsky at the Pharmacology Department of Harvard Medical School doing nuclear magnetic resonance (NMR) analyses of peptides proved disappointing to Bullock. In those 60-megahertz, horse-and-buggy days of NMR instrumentation, the turning on of the lights for night baseball at Fenway Park eliminated the power source for their experiments. Beyond the laboratory, the academic scene, caught from glimpses of the pressure to publish, grubbing for grants, and jockeying for tenure, seemed less than inspiring.

From Boston, Bullock went for a second fellowship year to work with Melvin Calvin in his multidisciplinary laboratory, the "round-house" on the Berkeley campus of the University of California. Physicists, chemists, and biologists were trying to understand the molecular behavior of charge-transfer complexes in photochemical model systems of photosynthesis. It was an exciting year, but the prospects of a top-line university appointment with a decent salary seemed remote. At Calvin's urging, he explored jobs in industry. Fieser offered him an introduction to Arthur D. Little, Inc., the premier research consulting company, which was loaded with 400 "grey foxes"—Harvard and MIT graduates, Fieser's students prominent among them. At age 28, with a wife and child to support, a salary increase from $5,500 to $14,500 was welcome, and keeping up with the swift scientific pace was a challenge.

Bullock's seven-year stay at Arthur D. Little was a transforming experience in learning how interesting, and how important, making business out of science and technology could be. In the company of M.B.A.'s and engineers, as well as chemists and biochemists, Bullock donned several hats. He prepared research proposals for industrial organizations and government agencies and took on other unsolicited projects. As the "case leader," he developed novel analyses for marijuana, antimalarials, and other drugs; methods were fashioned for measuring the toxic products in cigarette smoke and in spacecraft environments. The metabolites of these drugs and toxins in body fluids and tissues also had to be isolated, characterized, and quantified.

When a large Army contract on chemical warfare agents was cancelled, discussions began on starting a new pharmaceutical company, to be called SHARPS (an acronym based on John Sheehan [of penicillin fame], Lew Harris, Raj Razdan, and Henry Pars). An outgrowth of negotiations with Abbott led indirectly to an offer to Bullock to join their Life Sciences Division.

Starting at Abbott in 1972 as a chemist and group leader in gastroenterology, Bullock went on to become the head of the Metabolic Research Laboratory and then manager of Medicinal Chemistry in Research and Development of the Pharmaceutical Products Division. By the time he joined Schering-Plough in 1979, he had given up bench work and was directing about fifty synthetic organic chemists and biochemists, who dealt with bioassays, protein isolations, and chemical syntheses in the development of drugs for diseases of the cardiovascular, central nervous, and digestive systems.

Advice from DNAX directors and founders to Schering managers that they establish immunology and molecular-biology groups in New Jersey and recruit topflight people to work with DNAX scientists was initially rejected. "Let the East be East and the West be West," we were told. Eventually, our pleas to Luciano, Kogan, D'Andrade, and Lane were heeded, and a major reorganization was effected. Waitz would now report directly to Alex Lane, who then spent a microsabbatical at DNAX to become more familiar with the staff and the science. Nagabhushan was put in charge of all biotechnology at Schering-Plough, with the specific assignment to expedite DNAX discoveries for development and production. Bullock, relieved of direct control of DNAX operations, remained on the scientific advisory and policy boards and stayed intimately involved in implementing the Schering development of DNAX discoveries.

More recently, an IL-4/IL-10 team was organized. It includes scientists and clinicians drawn from all disease units at Schering-Plough, along with contingents from DNAX and Nagabhusban's group. They meet monthly to plan and coordinate strategy for the manufacture, clinical trials, and regulatory filings of these two cytokines. Considerable optimism prevails that one or another of the DNAX cytokines will prove to be an important drug in some aspect of cancer, septic shock, allergy, autoimmune diseases, or chronic bowel disease.

The DNAX Founders

"We expected you guys [DNAX founders] to lose interest and to move on to other ventures after a few years," intoned a top Schering-Plough executive. But the DNAX founders not only stayed but strengthened their personal and scientific ties both to DNAX and to Schering-Plough. Once, at an early stage, I had asked Hugh D'Andrade's permission to join the board of directors of another biotech venture. He demurred, not out of con-

The author (left), with Hugh McDevitt and Spyros Andreopoulos, at DNAX in 1985

Robert Luciano (left), Allan Waitz, and the author

cern for a conflict of interest but, rather, wanting my commitment to Schering-Plough to remain undiluted. He more than matched the financial increment the proposed directorship would have brought, and he extended it to the other Stanford founders. Later, with the growth in the stature of DNAX and in the strength of the Schering-Plough connection, our associations with other biotech ventures were welcomed as ways of extending our awareness in biotechnology and as possible opportunities for future interactions with Schering-Plough.

Financial inducements to the three Stanford founders were modest and muted; they were never discussed among the three of us nor with Zaffaroni. We felt compensated to be part of an exciting venture to advance the science we helped create and to apply it in a prompt and effective way. We three were given equal shares; only for this writing did I calculate that the value of each share was near two percent of the DNAX stock acquired by Schering-Plough. Since the acquisition, Schering-Plough has been most generous in its compensation for our continued service to DNAX.

Alex Zaffaroni, despite responsibilities at ALZA, an intense focus on the start of Affymax, and a weakening grasp of the growing complexity of techniques and terminology in molecular immunology, remained closely identified with DNAX. He could not attend all of the monthly executive committee and quarterly policy board meetings, but, for Berg, Yanofsky, and me, these dates held first priority on the calendar. According to some observers at DNAX, Schering-Plough, and elsewhere, the devotion of the Stanford three has been vital to the success of the DNAX–Schering-Plough connection. Our scientific advice has diminished with the sophistication of molecular immunology and the increased emphasis on it at DNAX, but we have remained responsive to the quality of the research, have praised personal leadership, and have encouraged communal focus. We have been vocal in the maintenance of principles and high standards, and have communicated our loyalty and devotion in the recruiting of scientists, staff, and advisors. Most important, perhaps, we conveyed to the Schering-Plough management our frank appraisals of people and progress at DNAX and of the problems we perceived in the DNAX–Schering-Plough connection.

Paul Schimmel, Professor of Biology at MIT and a member of the board of directors and the scientific advisory boards of Repligen Corporation, Alkermes, Inc., and Cubist Pharmaceuticals, Inc., commented in a review of an early draft of this book for the publisher:

> My sense is that Arthur, Paul, and Charley were among the few people who had the respect of both cultures and that, when things got tough, it was their commitment and involvement which kept the scientists from leaving and kept the investors from pulling the plug. Thus, what the book may need to develop better is the incalculable significance of a long-term commitment from scientific founders/directors/advisors who have the respect of the two cultures, who are willing to refrain from short-term financial opportunism, and who do not greatly change their lifestyle as a result of new-found wealth. The contribution of these individuals goes beyond establishing credibility in the beginning stages. They become substance and glue as well. When this kind of substance and glue is in place, you then have a success like DNAX. Although this message is implicit in the book, the broader lessons and generalizations are not emphasized, so that the message seems too much like a success story unique to DNAX.

What in the backgrounds of the "Stanford Three" had previously discouraged them from industrial associations, and what then happened in the DNAX–Schering-Plough connection that made them as devoted to the DNAX enterprise as to their treasured academic niches? The paths Berg and Yanofsky took to this venture follow; mine was described in Chapter 2.

Berg, Yanofsky, and I share a similar cultural heritage. We were raised in poor to modest circumstances in New York, we were educated in its public schools, and we were expected by indulgent parents to excel in our studies. There were no traditions of professional scholarship and certainly no particular knowledge or appreciation of science. We all ended up with an intensive dedication to careers in biochemical research. Yet we arrived there having taken different paths and with distinctly different scientific and personal styles.

Paul Berg, inspired to do medical research, gravitated to biochemistry, where his curiosity, energy, and creativity quickly brought him to the forefront of the enzymology that seeded molecular biology and ignited the explosion of genetic engineering.

Paul Berg

Paul Berg, who was born in 1926, grew up in the southern part of Brooklyn, in a residential enclave called Sea Gate, next to the major bathing beach, boardwalk, and amusement parks of Coney Island. An interest in science aroused by reading *Arrowsmith* and *Microbe Hunters* was further stimulated in the nearby Abraham Lincoln High School. The excitement conveyed by qualified and devoted science teachers in the New York City secondary schools fifty years ago far exceeded what is generally available today. Bright and eager students were encouraged to search for answers by reading and, on occasion, by doing experiments.

At age 16, Berg enrolled at New York's elite City College as a chemical engineering major. Bewildered by a curriculum he never anticipated, forced to register for courses given at awkward hours, faced with a two-hour commute each way on crowded subways, and alienated by highly competitive classmates, Berg quit after three days. He switched to the nearer (and friendlier) Brooklyn College and to a major in biology, which was nearer to his interests.

Leafing through a course catalog of Pennsylvania State College (later University) that had been supplied by a friend attending the college, he discovered the existence of biochemistry, a subject even closer to what he wanted to do. With savings from summer jobs, he went west to the lovely campus in central Pennsylvania, with its easy registration, its relaxed atmosphere, and its tuition of $100 per year. Some months later, when Berg's enlistment in the Navy was activated, he continued to attend the same classes, but then in uniform, and with all expenses paid. Halfway through, he was ordered to midshipman's school, from which he graduated in June 1945 with the rank of ensign. Assigned to a subchaser and perpetually seasick, he was still disappointed, two months later, that the defeat of Japan meant that he would not see active combat.

Berg left the service in 1946 and returned to Penn State to complete the last two years of his biochemistry major. He wrote out more than a hundred applications for summer jobs and landed one in the research laboratories of General Foods in 1947 and, the next summer, one with the Lipton Tea Company. He found the analytical and preparative chemistry enjoyable and similar to the nutritional, blood, and urine biochemistry he had been taught at Penn State. He had little awareness of the

Paul Berg

new excitement about enzymology and energetics, but he
caught a glimmer of this other world in reading the literature in
preparation for the writing of a senior paper. One of the published reports revealed that biochemists at Western Reserve
University (which later merged with the Case Institute of Technology to become Case Western Reserve University) in Cleveland were using isotopic tracers to discover fascinating aspects
of the intermediary metabolism of cells and organisms. Berg
also realized, from his summer employment, that it was the
Ph.D. who got to be a group leader in industry, whereas those
with only a B.S. were consigned to performing routine tasks.

He applied to many Ph.D. programs in biochemistry. The
one at Western Reserve, the one he wanted most, turned him
down; an offer from Oklahoma A. & M. (now Oklahoma State
University), which also included a job for his wife Millie, was
the best of the remaining possibilities. They had been married a
year earlier when Millie had finished nurse's training in Brooklyn, but, for lack of housing or a job for her at Penn State, they
had been forced to live apart. Then, a late acceptance from

Western Reserve and the gracious agreement by Oklahoma to let them change their minds, propelled the Bergs on a course to Cleveland.

The Western Reserve fellowship was not in the Biochemistry Department proper but in a clinical cul-de-sac remote from the center of the basic research. Under the direction of Leonard Skeggs, an innovator in automated clinical chemical analysis and in renal dialysis, Berg's job was to process huge volumes of human urine for the concentration and isolation of urokinase, thromboplastic factor, and other biologically active factors, which were then assayed laboriously in dogs. After two years of this kind of work, a seminar he gave in a graduate course caught the attention of Warwick Sakami and other members of the biochemistry faculty and brought an invitation to switch to the graduate program of the main Biochemistry Department.

Berg was influenced by G. R. (Bob) Greenberg, who was using tracers labeled with carbon-14 (a radioactive isotope) in cell extracts to define the pathway by which the purine building blocks of the coenzymes and nucleic acids were made. He tried a similar approach to explore the source of the methyl group in methionine. The established dogma at the time held that human tissues were unable to synthesize the methyl group, a deficiency that explained why methionine and choline were essential nutrients in the diet. He incubated formate or formaldehyde, each labeled with carbon-14, with liver extracts, and then carried out an elaborate chain of degradative steps to recover the methyl group from methionine. Remarkably, these one-carbon compounds were found to be effective precursors of the methyl group. Later on, nutritional studies confirmed that methionine can be made in the body and is not essential in the diet as long as vitamin B_{12}, folic acid, and homocysteine are consumed in adequate quantities.

These findings, presented at a national scientific meeting, attracted the attention of Vincent du Vigneaud, a major figure in biochemistry and the authority on methyl groups. Berg's responses to vigorous questioning impressed du Vigneaud so much that he approached Harland G. Wood, who was then the chairman of the Biochemistry Department at Western Reserve, with an offer to Berg of an assistant professorship at the Cornell University Medical School, only to be informed that he was still a graduate student.

When Berg finished his Ph.D. thesis in June 1952, he was eager to do postdoctoral work in enzymology, and he chose to work with Herman Kalckar for a year at the University Institute of Cytophysiology in Copenhagen and then with me at the NIH in Bethesda. Harland Wood approved of the Kalckar choice but insisted that Berg spend his second year (1953-1954) with the eminent Carl Cori at Washington University in St. Louis. Paul resisted. Despite the prestige of the Cori laboratory, he could not overcome an aversion to living in St. Louis. How ironic it was that, in the fall of 1952, I wrote to Paul, already in Copenhagen, that I had accepted a position at Washington University and would be moving to St. Louis the next year. Three other gifted postdoctoral applicants that year, when informed I was moving, sought out other advisors at the NIH, but Berg stuck with me and was willing to go to St. Louis after all.

In Copenhagen, Berg discovered nucleoside diphosphate kinase, the all-purpose enzyme that converts a nucleic acid precursor to the active triphosphate form. In St. Louis, he made the even more impressive discoveries of how acetic acid is activated on the way to making lipids and steroids. The mechanism of acetic acid conversion to the coenzyme A form, one of the key intermediates in metabolism, was then at center stage in biochemistry. This finding had been sought by Fritz A. Lipmann, who was awarded the Nobel Prize in 1953 for the discovery of coenzyme A, and by Feodor Lynen, whose clarification of the mechanism of cholesterol biosynthesis had earned him Nobel Prize recognition in 1964. In addition, Berg discovered how an amino acid, such as methionine, is readied for assembly into proteins. His studies of methionine activation launched a decade of work that kept him at the forefront of protein synthesis.

Over the years, Berg's devotion to enzymology became diluted by an interest in microbial genetics, which had been developed in after-tennis discussions with Yanofsky. Work with Yanofsky on how suppressor genes overcome mutations, and a long exposure to the studies by Dale Kaiser of a bacterial virus called lambda, impressed him with the power of genetics and the utility of bacterial viruses. This led Berg to wonder whether lessons from the latent state of a bacterial virus might be applied to understanding how the integration of a tumor virus transforms an animal cell so that it exhibits uncontrolled growth. With this objective, he spent a sabbatical year in Renato Dulbecco's lab at the Salk Institute to learn the techniques of animal-cell culture in order to work with the polyoma and simian (SV40) tumor viruses.

Could SV40 DNA, which can enter animal cells, become a ve-
hicle for introducing a bacterial gene whose expression could be
readily monitored and controlled? This would require that the foreign gene be spliced into the SV40 DNA without interrupting the
viral genes needed for replicating the DNA. In a reciprocal way, if
means could be found to introduce DNA into *Escherichia coli*, then
the SV40 tumor genes might be expressed in the bacterial cell.
Again, the foreign tumor gene would have to be spliced into the
bacterial DNA, such as that of a plasmid, in a way that would not
interfere with the plasmid's capacity to sustain its own replication.

These splicing techniques for generating recombinant DNA
were discovered in the Stanford Biochemistry Department by Berg
and his students and by Peter Lobban, a graduate student with
Dale Kaiser. They used enzymes that cut, fill, and seal breaks in
DNA, enzymes that my group and the group of my close colleague
Robert Lehman had discovered some years earlier. The generation
of recombinant DNA and the introduction of foreign genes into
bacterial and animal cells depended on two other technical advances made by others. Enzymes called restriction nucleases had
been discovered that cut DNA at precise places and afforded the
means for splicing genes together. About the same time, a procedure was perfected for breaching the *E. coli*'s cell-membrane barrier to DNA entry. Paul's large share in these discoveries was recognized in 1980 by the Nobel Prize in Chemistry, which he shared
with Walter Gilbert and Frederick Sanger, who had perfected techniques for determining the sequential order of nucleotides in DNA.

The ability to modify bacterial and viral species at will, and thus
to create bacterial and plant factories expressing foreign genes, set
off major societal reactions in which Berg took an active role. One
concern was that genetic engineering might run amok and generate environmental disasters. A moratorium on all such genetic
research was declared, and a meeting was convened, in 1974, at Asilomar, a conference center on the California coast, to consider the
issue. As chairman of the conference, Berg took a leading role in
formulating guidelines for future research and served as spokes-
man for the adopted recommendations. His interest and skill in in-
terpreting science to disparate groups were clearly manifested.
These qualities have surfaced many times since, notably in the recent organization and funding of the Beckman Center for Molecular and Genetic Medicine at Stanford, of which he is the director.

Twenty years after Asilomar, and after the most extensive use of recombinant-DNA technology imaginable, not a single untoward result of genetic engineering has been recorded. On the other hand, the commercial exploitation of genetic engineering has borne enormous benefits, and it offers even more for the future. Venture capitalists fueled the explosion of biotech ventures. Their initiative enlisted increasing numbers of biologists and chemists familiar with recombinant-DNA technology and its possible application to medicine and agriculture. For some, the motivation was simply to make money—and in a hurry. Their concerns were for profitable products and not for increased knowledge. They regarded scientists, like Berg, whose work had made their business possible, with indifference, or with the disdain for "those who were left behind." Chafed by an attitude that was superficially polite but basically unfeeling, Berg was eager to join with trusted friends in the creation of DNAX, an enterprise in which topflight science linked to long-range medical objectives, would be an attractive entry into the biotech parade.

Charles Yanofsky is one of the very few scientists who is genuinely bilingual in biochemistry and genetics. He grew up that way. Biochemists who discover the power and practice of genetics later in life reveal the accents of their first language. Geneticists who later turn to biochemistry are also marked by the foreign intonation of their mother tongue. Not so Charley.

Charles Yanofsky

Charley Yanofsky was a graduate of the remarkable Bronx High School of Science in 1942, soon after its founding. The school was a short walk from where he lived. His experiments in high school included an original study of ultraviolet-induced mutations in fruit flies (*Drosophila* species). At the City College of New York, he was a biochemistry major, but he still retained his interest in genetics. Midway through college, he was drafted into the Army as a private, and he saw combat in the Battle of the Bulge. Mustered out as a corporal in 1946, he returned to City College to complete his B.S. degree. He had sought admission to a number of colleges outside New York City, but anti-semitic prejudice apparently outweighed his scholastic record.

As a senior in college, he was inspired by the "one gene, one enzyme" revelations of George W. Beadle and Edward L. Tatum, who, for this discovery, shared, with Joshua Lederberg, the Nobel Prize in Physiology or Medicine in 1958. He was determined to do his graduate work with one of them and was happy to accept Tatum's offer at Yale because Carol, his fiancée, needed another year to finish her studies at Brooklyn College. (He had been turned down by Beadle at far-away Caltech.) But when he arrived at Yale in 1948, Tatum had just left to return to the Biological Sciences Department at Stanford.

At Yale, Yanofsky did his thesis work with David Bonner, who had taken over Tatum's group. A lively iconoclast who was denied a faculty appointment for some years, Bonner was a devoted friend and teacher. He shared with his students his acquaintances with the great geneticists of the generation by inviting them to visit and to teach guest courses at Yale, such as a memorable one on the single "R" locus of maize given by the venerated corn geneticist Louis Stadler. To be excused from the required biochemistry course at Yale, Yanofsky informed Joseph Fruton, the chairman of the department, that he already knew everything in Fruton's authoritative textbook, whereupon he was required to undergo a rigorous tutorial on the classic biochemical literature that only the highly critical Fruton could administer.

Yanofsky's graduate research on the pathway of niacin and tryptophan biosynthesis in *Neurospora crassa*, extended by three years of postgraduate work on these pathways in *Escherichia coli*, set the pattern for his life's work, the comprehensive and incisive explorations of the biochemical, genetic, and physiological features of metabolic operations in a free-living cell. Upon moving to the Western Reserve University School of Medicine in Cleveland in 1954, as an assistant professor, he brought to the Department of Microbiology his unique expertise in microbial genetics. The remarkable synergy of that department with Harland Wood's outstanding Biochemistry Department strengthened his grasp of enzymology and intermediary metabolism.

After four years as an assistant professor, Yanofsky was offered several positions, including an associate professorship in Biological Sciences at Stanford. Tatum had just moved on to Rockefeller University, leaving Stanford a sleepy place in biology. But President Wallace Sterling and Provost Frederick Terman, determined to revive first-class science at Stanford in all departments, persuaded Yanofsky to accept the Stanford offer.

Charles Yanofsky (second from right), with Paul Berg, Satwan Nerula, and Kenichi Arai

How strange that, ten years after seeking Tatum at Yale, Yanof-sky would again find Tatum's trail and this time inherit his laboratories and hand-blown glassware in the quaint catacombs of Jordan Hall.

Sterling and Terman had engineered the move of the Stanford Medical School from San Francisco to the Palo Alto campus and had recruited me and my department from Washington University in St. Louis to organize a proper biochemistry department. With this development, Joshua Lederberg felt persuaded to leave the University of Wisconsin to found the Department of Genetics in the medical school. All the while, bold plans were also afoot to revamp Stanford's chemistry department. On his way for the interviews at Stanford, Yanofsky stopped in St. Louis, stayed with Paul and Millie Berg, met our group, and was caught up in our excitement to leave the Gateway to the West for the promised land.

With his talents to inspire and direct young people and his technical skills to improvise and to use genetic and biochemical strategies, Yanofsky continued his unwavering exploration of how *E. coli* manages the frugal use of tryptophan and its efficient biosynthesis. From these studies emerged revelations of the complex network of enzymes, promoters, and repressors, the stratagems of an awesome economy that makes the *E. coli* cell competitive in feast or famine, in growth or quiescence, and under stress of every kind—all with the most sparing use of external resources. In the course of this focused study, Yanofsky discovered some of the key principles of gene structure, function, and regulation:

- He established the colinear relationship of the nucleotide sequence of a gene to the amino acid sequence of the proteins it specifies, thereby verifying one of the fundamental postulates of genetic chemistry and laying the foundation for the solution of the genetic code.

- He postulated, and then proved, that missense mutations can be suppressed by translational misreading of the mutant codons by altered molecules of transfer RNA (tRNA).

- He determined the chemical structure of the five genes and the five enzymes of tryptophan biosynthesis and identified the precise chemical interactions between a regulatory protein and a nucleic acid sequence that regulate the output of that set of genes.

• By discovering a then novel regulatory mechanism for attenuating gene expression, he showed how the cell can sense the level of the tRNA charged with tryptophan and use this information to determine whether transcription, once initiated, produces translatable messenger RNA (mRNA) or is terminated before reaching the genes that encode the enzymes that manufacture tryptophan. His chemical and genetic analysis of this mode of regulation revealed a mechanism of incredible simplicity, beauty, and significance: alternative structures in the nascent messenger RNA that influence the ability of the ribosomes (the cell's protein-synthesizing machines) to translate the DNA code word for tryptophan.

These findings, reported in a magisterial succession of papers and lectures over more than 40 years, are models of clarity and understatement. Reserved by nature, self-confident yet modest, Yanofsky became the scientist's scientist.

For his profound discoveries and for his influence on an entire generation of molecular and cellular biologists, he has received many prestigious awards—presidencies of both the Genetics Society of America (1969) and the American Society of Biological Chemists (1984), membership in the National Academy of Sciences (1966), Foreign Fellowship in the Royal Society (1985), and a dozen other major prizes and honorary degrees. But, the Nobel Prize has thus far been denied him, as it has to some others who richly deserve it. Neither in him nor in some others I've known well, who, like him, have not received this deserved recognition, did I ever detect disappointment or resentment. What matters most to Charley is the understanding that comes from doing good science. As for rewards, he feels fulfilled by the approbation of his peers in genetics and biochemistry and by his selection as a Career Investigator of the American Heart Association.

The Career Investigatorship confers immunity from certain administrative activities, such as the chairmanship of an academic department, but it has not diminished Charley's involvement in teaching nor has it limited his responsibilities to advisory groups both on and off campus. Nor, for that matter, did it enjoin his participation as a founding member of DNAX.

In earlier encounters with pharmaceutical companies, Charley was disappointed by short-term objectives and by a general reluctance to probe the basic aspects of a problem. Overtures from venture capitalists to develop and apply the techniques of genetic engineering, in which he was a leader, left him even more uneasy with their aggressive timetables for progress and their intense focus on profits. Yanofsky discovered, as Berg and I had found, that DNAX, under the guidance of Alex Zaffaroni, offered an agreeable avenue to advance technology and science with close friends and kindred spirits.

The DNAX Future

Cytokine research—the main line of DNAX investigations—can be seen in stages (Figure 13), from the cloning of genes for novel cytokines, determining their effects on a variety of cell types in cultures, and culminating, at the end of the 1980s, in the discovery of the TH1 and TH2 (helper T cell) paradigm and the remarkable cytokine IL-10. This research set in motion current attempts to find the uses to which cytokines are put in the animal body, particularly by means of gene knockouts in mice. These animal studies will continue for much of the 1990s because they reveal extremely useful basic information and offer direction to Schering-Plough for promising avenues of clinical development.

Cloning the genes for the cytokine receptors and determining the structure of the cytokine–receptor complexes are essential early steps in outlining the pathways by which the signals generated by the cytokines are transmitted to the DNA in the nucleus. Identifying the molecular basis for cytokine action from the cell surface to the gene will provide Schering-Plough with assays for drugs to potentiate or interrupt the process, depending on the disease in question. Innovative approaches are being applied to the discovery of many novel genes in the responses of tissues to immunological, inflammatory, and hormonal stimulation. These directions of research inside and outside the cell should keep DNAX at the major frontiers of science, extending beyond the immune and blood-forming systems to aspects of growth and development and the aberrations of cancer, degenerative diseases, and aging.

In the administrative sphere, some changes might have been expected with Al Waitz's retirement in 1992. He could reflect on ten

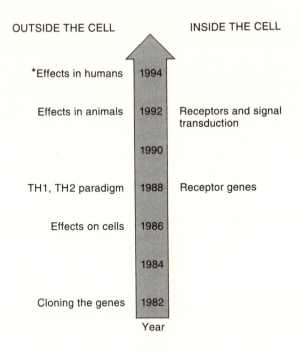

OUTSIDE THE CELL   INSIDE THE CELL

*Effects in humans	1994	
Effects in animals	1992	Receptors and signal transduction
	1990	
TH1, TH2 paradigm	1988	Receptor genes
Effects on cells	1986	
	1984	
Cloning the genes	1982	

Year

*Development at Schering-Plough

FIGURE 13. Stages in cytokine research at DNAX.

years of successful blending of the cultures and operations of the industrial East and academic West. But 25 years of the rough-and-tumble in the pharmaceutical industry had taken a toll. He was drawn to an outdoor life of fishing, gardening, and building. Managing his 70-acre ranch near Eugene, Oregon, close to his favorite trout streams, with occasional consulting with biotech companies, has so far been the right recipe for a happy retirement.

In seeking Waitz's replacement, the advice to us from Schering-Plough top management was, "Keep the DNAX ship on course." We agreed, mindful that squalls, storms, and crew changes were inevitable. We needed a captain who could adjust the course to keep the ship steady as she goes, and we found the right person in Jacques Chiller.

Jacques Chiller

Born in Paris in 1936 to Polish immigrant parents, Jacques
Chiller migrated with his family to Vancouver, British Colum-
bia, when he was 11. He finished high school in Bellingham,
Washington, and he went on to the University of Washington in
Seattle as a predental student, but without any strong career
direction. Emerging in 1962 from a two-year stint in the U.S.
Army, he did odd jobs before returning to college, where he
discovered the excitement of immunology and biochemistry. A
Ph.D. thesis on comparative immunology was followed by post-
doctoral work on immune tolerance at the Scripps Clinic
Research Institute in La Jolla at the explosive time when T and
B lymphocytes were being distinguished and cellular immunol-
ogy was in its robust infancy.

After professorial appointments to the National Jewish
Hospital and Research Center attached to the University of
Colorado Medical School, Chiller took a sabbatical at the Immunology Institute in Basel. It was then that Frank Dixon, Director
of Scripps, prevailed on Chiller to return to La Jolla to forge
and direct an industrial relationship between the Eli Lilly Company and Scripps. Only two years later, in 1983, this would be
superceded by a more extensive arrangement between Scripps
and Johnson & Johnson. Lilly then decided to establish its own
research facility in La Jolla and installed Chiller as its director,
with a mandate for a continuing focus on T-cell cytokines and
their regulation.

In 1987, when Lilly acquired Hybritech, with its emphasis on
diagnostics and imaging procedures, Chiller was asked to wear an
additional hat as its senior vice president. Out of loyalty to Lilly, and
with a curiosity about the technology and business side of an enterprise, Chiller in this new role became aware of his talents as an
organizer and of the pleasure of cheering others on to their success.
It was a turning point from laboratory investigation, and it precip-
itated a move, in 1988, to the central Lilly Research Laboratories
in Indianapolis, where he became Vice President for Research, replacing the redoubtable Irving Johnson on his retirement. In
charge of more than a thousand people in the drug-discovery divisions, he vigorously recruited highly placed academic leaders to

Jacques Chiller

come to Lilly with a vision to invigorate Lilly research for the next century.

Then the call came from DNAX. It was irresistible. Although it would be wrenching to leave the people he had persuaded to join him at Lilly and to forgo participating in the fruition of their plans, his love for science was far greater than for management. Chiller's move from Lilly to DNAX would mean diminished administrative responsibility, but he saw an opportunity to return to a more academic setting, to revive an intimacy with science and immunology, and to enjoy the California climate and style that he had missed in Indianapolis. Chiller brought an enthusiasm to maintain the high level of science at DNAX and the conviction that he could enhance the vigor of the Schering-Plough connection, which would require widening and strengthening the bridges between Palo Alto and New Jersey. As an immunologist, he felt secure in judgments about the directions and people at DNAX. To Schering-Plough, he would bring his experience in clinical applications and company management to bear on improving the quality of their drug research and the development of DNAX discoveries into marketable products.

What of the future? When the clock strikes midnight to begin 1 January of the year 2001, there will surely be an emotional response to the dawn of a new millennium, only the third in the combined histories of the Julian and Gregorian calendars. Anxieties will be felt in boardrooms, laboratories, and households throughout the world whether the year 2001 will usher in decades better or worse than those that preceded them. How will DNAX and Schering-Plough be prepared to face that future?

At DNAX, we can hope that the administrative and scientific leadership will remain stable and will continue to mature. In order to resist growth in size, the changes (which are so essential for progress in science) must come from some turnover in the senior staff, maturation of the junior staff, and a constant flow of postdoctoral fellows. Will that be adequate to keep DNAX vibrant, adventurous, and at the forefront of its selected areas of research? Might the lack of challenge from students, teaching, and the tensions of competition for research grants and fellowship support lead to complacency and stagnation?

At Schering-Plough, the future has other uncertainties. There will be major changes in the leadership of the company. Alex Lane, Chairman of the Schering-Plough Research Institute, and Frank Bullock, the senior vice president in charge of research operations, retired in 1993, and Bob Luciano plans to step down as CEO in 1995. Only Hugh D'Andrade will remain in his accustomed role of strategic planner. Dick Kogan, who is succeeding Luciano, brings superb business acumen and an understanding of the biotech impact on pharmaceutical advances. Jonathan Spicehandler, in replacing Alex Lane, brings experience in clinical research that will need to be expanded to other levels of the discovery effort. There, he will have the support of Cecil B. Pickett, newly recruited (from his position as senior vice president for basic research at the Merck Research Laboratories in Montreal) to replace Frank Bullock. DNAX was a major attraction for Pickett in his decision to join Schering-Plough; effective interactions between him and Jacques Chiller can be expected. Barring dramatic setbacks in Schering-Plough's business, there is every likelihood of continued support for DNAX.

Most important for the future of the DNAX–Schering-Plough connection is the leadership of Richard Kogan. With his background in finance and his superb business sense, he has helped steer the profitable growth of the company since arriving in 1982. He will determine the directions of all corporate activities, including the emphasis on research and on DNAX's place in it.

Dick Kogan, at his first board meeting, which came within days after joining Schering-Plough, heard the presentation that Alex Zaffaroni made for the deal with DNAX. He was as impressed by Alex then as he had been when he had served as a Ciba-Geigy representative on Alex's ALZA board. Soft-spoken but intense, Alex was persuasive in his pleas for long-term investments to "create value." Although Luciano had made up his mind to close the DNAX deal, he needed the board's support to pay nearly $30 million for what may have seemed like little more than an idea, especially so early in his tenure as CEO, and at a time of sharply declining profits for Schering-Plough.

Kogan and Luciano share many similarities of background. Separated by eight years in age, both grew up in the Bronx in poor circumstances, went to Christopher Columbus High School, and then attended City College. Neither showed any early inclination toward science, yet both went on to manage pharmaceutical companies, and both had long assignments at Ciba-Geigy before teaming up at Schering-Plough.

Richard Kogan

During his school years, Richard Kogan, then small and tough, tended bar in his father's bar and grill on Tenth Avenue in Manhattan's infamous Hell's Kitchen and hawked beer, peanuts, and popcorn in the stands at Yankee Stadium and at the old Polo Grounds. He acquired both street smarts and good enough grades for admission both to Columbia University and to tuition-free City College. He chose the latter because of his father's promise of a car in lieu of the Columbia tuition, but, as it happened, he never did get the car.

At the uptown campus of the still-prestigious City College, his major in economics focused on the mathematical side of the subject. He flirted with a Ph.D. program at Rutgers with an academic career in view, but he soon quit out of boredom and

Richard Kogan

uncertainty. Subject to the draft, he spent time on active duty and in the reserves, rising from private to sergeant. He could then have taken a job with General Foods in White Plains, N.Y., but he chose Warner-Lambert, a pharmaceutical company, instead, because their location in Morristown, N.J., was close to where his wife (Susan, a chemist) was employed, at Lever Brothers in Edgewater, N.J.

After working with controllers and other financial people for four years and gaining an MBA from New York University on theoretical aspects of finance (with a thesis entitled *The Right Return on Investment*), he was ready to move from analysis to operations. In 1969, Don MacKinnon (who had gone to Ciba as chief financial officer from his Warner-Lambert post as treasurer) provided the opportunity. After two years at Ciba, Kogan was ready to take on successive directorial responsibilities: in chemicals and dyes, in mergers and acquisitions, and then in planning and administration, replacing Bob Luciano when he moved up the executive ladder.

As president of Ciba-Geigy's Pharmaceutical Division in Montreal from 1976 to 1979, Dick Kogan enjoyed the opportunity to manage a business on his own for the first time. When he returned to the U.S. headquarters as the president of the Pharmaceutical Division, he also represented Ciba-Geigy on the ALZA Board of Directors. In 1982, after thirteen years with Ciba-Geigy, he was happy to stay on, expecting either to move to Basel to head up the Pharmaceutical Division or to become the president of the American corporation. This stable, well-tuned plan was suddenly revised when Bob Luciano invited Kogan to be Executive Vice President of Pharmaceutical Operations at Schering-Plough, with the likelihood—but no assurance—of advancing to one of the top positions. In 1986, Kogan became President and Chief Operating Officer of the company. He also served as President of the Pharmaceutical Manufacturers Association during 1990–1991.

The top management at Schering-Plough—Bob Luciano, Dick Kogan, and Hugh D'Andrade—have worked effectively together for more than ten years and have known and trusted each other for more than twenty. Luciano is confident, quick, and intuitive; D'Andrade is logical, sharp, and deliberate; and Kogan is bright, resourceful, and savvy. They've made a great team at Schering-Plough and have provided DNAX with the consistent support that a basic research organization needs for attaining long-term objectives.

Kogan, like the chief executives of the other major American pharmaceutical companies, will be facing the prospect of severe price controls and increasingly costly programs of drug discovery and development. New health plans may leave little incentive for massive investments in new drugs, and they may diminish the capacity of pharmaceutical companies to withstand inevitable failures at stages of research, development, drug approval, or marketing. There is a real need for innovation to produce better drugs to cope with an unending succession of diseases. The challenge will be to provide greater novelty and efficiency in drug discovery by more and better science and to improve the integration of the discovery, development, and marketing arms of a pharmaceutical company.

At a meeting of the DNAX scientific advisory board in May 1994, three members—J. Michael Bishop of the University of California, San Francisco, Max Cooper of the University of Alabama, and Klaus Rajewsky of the University of Cologne—each noted the remarkably congenial sociology of the scientific staff and management, which was unique in their experience. Can these attitudes be maintained in the face of considerable turnover in personnel and changes in the climates of science and business?

The greatest uncertainty for DNAX lies in the future role performed by the founders—Paul Berg, Charles Yanofsky, Alex Zaffaroni, and me. In the year 2001, our ages will range from 75 to 83, ages that are decades removed from those of the new Schering-Plough managers and the DNAX staff. We hope that younger scientists will be found to further the ideals and foundations we laid at DNAX and that the ties to Schering-Plough will remain firm and fruitful for the next generation.

CHAPTER 7

Genentech, Amgen, Chiron, and Regeneron

E ach of the ventures described in this chapter and each of their founders deserve the space and attention devoted in this volume to DNAX and its creators, but I will present here only brief accounts to illustrate the diversity of origins and styles of biotech ventures. In different ways, each of these ventures has made significant contributions to science and has developed important products for the prevention or treatment of disease. Their successes have inspired new biotech startups—Tularik, Sugen, ICOS—which have borrowed heavily from the science and business management of the more established ventures. All these ventures, with distinguished founders, directors, and advisors, convey the excitement of expanding the reach of science beyond what can be initiated and pursued in conventional academic research and pharmaceutical company settings.

DNAX was started and molded by a group of senior scientists who maintained an academic research atmosphere with total support from a major pharmaceutical company. Genentech, "first on the block," was propelled by junior scientists without overt direction. They pioneered and focused the technologies of genetic engineering with a huge success that encouraged rapid business expansions, which briefly diverted the thrust of their science. Amgen, although with less notable scientific achievement, was managed well, developed two highly profitable products, and is becoming a fully integrated and independent pharmaceutical company. Chiron, driven by a triumvirate of scientists with organizational and scientific talents applied to vaccines and other molecules neglected by drug companies, emerged from the pack, swallowed

Cetus (the first of the biotech ventures), and is swimming strongly against the tide. Regeneron was organized and guided by leaders in molecular neuroscience to promote basic research in this field on diseases of the nervous system, as DNAX has in immunology; for lack of the long-range sponsorship enjoyed by DNAX, Regeneron has had to pursue the costly and diverting development of its own discoveries.

Genentech, Biotech Pioneer

Genentech was conceived in 1976 when Robert Swanson and Herbert Boyer decided to commercialize recombinant DNA. An undergraduate in chemistry at MIT, Swanson had gone on to business school and worked for Eugene Kleiner and Tom Perkins in their venture capital firm. A seed investment from the firm supported some of the research in Boyer's laboratory at the University of California, San Francisco (UCSF), and in the laboratories of Keiichi Itakura and Arthur Riggs at the City of Hope National Medical Center in Duarte, California. Combining the talents of both laboratories, they synthesized the gene for a human hormone, somatostatin, and attached appendages for cloning the protein in the intestinal bacterium *Escherichia coli*. With this success, Genentech, until then a company on paper, was emboldened to become a physical entity. A laboratory was outfitted in an old warehouse near San Francisco Bay in South San Francisco, where the nascent venture planned further to exploit the capacity of bacteria to become factories for the production of still other hormones, such as human insulin and human growth hormone, which could be put to immediate clinical use.

One of the best things that happened to Genentech was the signing of David Goeddel. Raised in Poway, California, and an undergraduate in physical chemistry at the University of California, San Diego, Dave intended to pursue his Ph.D. in physical chemistry at the University of Colorado in Boulder. Instead, he began working under Marvin Caruthers and was caught up in the organic synthesis of short pieces of DNA designed for assembly into full-length genes. By the time Dave finished his degree in 1977, he had also become familiar with the enzymes, plasmids, and cloning procedures then in use in recombinant DNA work. Without pausing for a postdoctoral fellowship, Dave took a job at the former Stanford

David Goeddel

Research Institute (SRI) in Menlo Park, California, to work with Dennis Kleid, who had received his training in DNA chemistry with Gobind Khorana and in molecular biology with Mark Ptashne.

Because research at SRI is funded by contracts, Kleid contacted Bob Swanson at Genentech for contractual support to synthesize genes. Instead of offering a contract, Swanson invited Kleid and Dave to join the infant company. Kleid vacillated at first, but Dave accepted promptly to become, in early 1978, the first full-time Genentech employee. But Dave was actually the second scientist to be hired at Genentech; Axel Ullrich, a German postdoctoral fellow at UCSF, had signed a contract, only to withdraw a few hours later with misgivings about taking an industrial job. The previous year, Swanson had tried to interest Axel, Peter Seeburg (also a German), and John Shine (an Australian), all three postdoctoral fellows at UCSF who were deeply engaged in recombinant DNA cloning, to join Genentech. He promised a lab and resources to continue their work, but their commitment to academic careers in their native countries seemed unshakable. Then, a year later, disenchanted by the squabbling over patents, publications, and authority among the scientists at UCSF, and with the lure of a setting in which Goeddel had cloned the synthetic DNA encoding the A and B chains of human insulin in just a few months, all three signed on. Shine returned soon to Australia, but Ullrich and Seeburg both stayed for ten years.

Ullrich, who had been an undergraduate at the University of Tübingen and had earned a Ph.D. in molecular biology under Professor Ekke K. F. Bautz at the University of Heidelberg, had come to UCSF for postdoctoral training, fascinated by the new recombinant DNA technologies. With Boyer's lab filled, he was shunted to Howard Goodman, another professor in the department working with recombinant DNA. Despite getting little help or encouragement, Ullrich was the first to clone rat insulin, at the same time that Seeburg, under similar circumstances, succeeded in cloning the rat growth hormone. Yet the omnibus patent for these discoveries was issued only to Goodman and William Rutter, who were in charge of the UCSF laboratories, excluding the young scientists whose initiative and work were responsible for the discoveries.

At Genentech, Ullrich and Seeburg were offered unlimited resources and an attractive ambiance for exploration and conquest. Their longing to return to academia in Germany was tempered by the realization that the major support and thrust to pursue the new science would be lacking there. The years 1978 through 1982 were golden ones in the Genentech laboratories, during which bright kindred spirits working furiously, energized by a team spirit, won many of the cloning races. The team was a unique combination of individuals challenged to do pioneering research, motivated by intense competition and by a sense of being outcasts from academic science. With exciting work spilling out in numerous papers published promptly in the elite journals, Genentech assumed the premier place in biotechnology.

The cloning of the genes for human and rat insulin is recounted engagingly in *Invisible Frontiers*, by Stephen S. Hall, which depicts the mad race won by Goeddel and Ullrich over Walter Gilbert, leading a Harvard team. The book is an excellent source, detailing the parts played by Boyer and Swanson, by the bench scientists and entrepreneurs, and by UCSF and Eli Lilly and Company in the early history of Genentech. The vivid characterizations of contestants in the cloning races—Goeddel and Gilbert, and the many others in supporting roles—are likely valid ones, and they portray a sorry picture of personal rivalries in the conduct of science. The point that scientists can be intensely competitive and even unscrupulous in pursuit of a laudable goal was made earlier in James Watson's account of the discovery of the structure of DNA in his 1968 best-seller, *The Double Helix*.

What is dismaying in Stephen Hall's account are his impressions, very widely shared, that the revolutionary advances of genetic engineering stem from scientific triumphs made by industrial and academic groups vying in the races to clone genes and to produce clinically relevant products. Utterly ignored are the decades of fundamental explorations of the chemistry of DNA, of the enzymes that assemble and rearrange DNA, of the genetics, plasmids, viruses, and physiology of microorganisms, and of the coalescence of all these basic studies and techniques in the creation of recombinant DNA. While the importance of applying this knowledge by Genentech and other biotech ventures to clone genes and to produce proteins for clinical and scientific use should not be minimized, neither should the scientific foundations on which this engineering is based be so completely disregarded.

Genentech boldly states, in its backgrounder handout to the media, that its "fundamental and most important asset is its science," and that "the company's research has helped lead to 7 of the 13 types of biotechnology-based pharmaceuticals on the market" (as of 1993). Genentech scientists lead industrial and academic departments worldwide in publishing, having produced 250 papers annually, including some of the most widely cited in the scientific literature. This thrust of Genentech, and its success, are due, in considerable measure, to Dave Goeddel.

Dave has been described by colleagues as a "kamikaze scientist," who works indefatigably day and night and is incredibly single-minded, driven, and competitive. He moved through the inevitable roadblocks with tank-like efficiency to clone the genes that made Genentech first among biotech ventures in papers and patents (currently, 1200 granted and 1100 pending). Lacking experience in biology and without exerting overt leadership, Dave still succeeded by virtue of his work ethic and his skills in motivating his gifted peers to emulate his productivity and purposefulness.

Having started Genentech, Herb Boyer was thereafter rather removed from the scene, except for his continuing influence on

Bob Swanson and the board of directors to maintain a strong focus on science and to encourage a flow of promptly published papers. Swanson, by contrast, was very much on the scene. In addition to managing the business side, he kept in close touch with scientific problems and progress and rallied the scientists almost daily to keep ahead of the competition in their cloning races.

Unlike other biotech ventures, with a seasoned scientist or a distinguished board of scientific advisors for guidance, Genentech relied on its "young Turks," unheralded but talented, industrious, and highly motivated to succeed. Notable among them were Ullrich, Seeburg, and Arthur Levinson. Ullrich and Seeburg brought considerable experience from their cloning of the insulin and growth hormone genes at UCSF, as well as biological insights to guide the technical applications. After ten years at Genentech, with scientific stature built from a wealth of significant publications, they were enticed by attractive academic offers to return to Germany— Ullrich, to assume directorship of the Molecular Biology Department of the distinguished Max Planck Institute of Biochemistry in Munich, and Seeburg, to a professorship and directorship of the Zentrum für Molekular Biologie in Heidelberg (ZMBH). Levinson built on his training in mammalian cell systems and gained the maturity to become the scientific director of the Genentech research enterprise of nearly a thousand scientists.

Upon announcing the cloning of the human insulin gene, Genentech was fortunate to have the endorsement of Eli Lilly, the company that had dominated the world market for porcine and bovine insulin for many decades and was highly respected in pharmaceutical circles. In addition to the immediate financial assistance that Lilly provided, the Lilly imprimatur was of incalculable value to Genentech in gaining the confidence of the investment community. The Lilly connection was crucial in still another way. Through the sagacious leadership of Irving Johnson, its scientific director, Lilly rapidly improved the primitive Genentech expression systems to achieve commercially viable levels. Lilly also provided the manufacturing facilities for a quality drug, and applied the regulatory and related expertise that, in 1982, brought the first recombinant DNA drug to market. Without Lilly's help, none of the resources essential for this integrated achievement could have been created at Genentech in a reasonable length of time.

After the cloning of human insulin in 1978 and of human growth hormone in 1979, Genentech made its initial public stock offering in 1980 and raised $35 million. Within an hour after the market opened, the price had risen from $35 a share to $89, one of the largest run-ups ever seen. The chagrin of those at Genentech for having so underpriced the stock was assuaged over the years by the image it built for the company as a premier investment. By contrast, Cetus Corporation, riding the crest of the Genentech wave, raised more than $100 million in 1981 with far less scientific substance. Poorly managed, Cetus succumbed to a buy-out by Chiron in 1992.

The impact of Genentech's success in the ensuing years was felt both in academic and in industrial circles. The excellent quality (and the large volume) of papers published promptly in the leading journals helped to erase the stigma attached to research careers in an industrial environment. The power of bright and energetic scientists to apply this new knowledge at a frontier of medicine excited investors and the pharmaceutical world.

The very rapid pace of success eventually overwhelmed Genentech. Even before the dashed hopes for billion-dollar sales of Activase (TPA, or tissue plasminogen activator), disaffections had set in. Some cloning races were lost—those for alpha interferon and erythropoietin. More seriously, Genentech's rapid expansion, particularly on the business side, introduced an alien culture without strong scientific leadership to counter it. In 1983, David Martin, Professor of Medicine at UCSF, with broad experience in biochemistry as well as in clinical medicine, was brought in to provide the scientific direction. However, the strong tides of product development, the large scientific staff, and the reluctance of Bob Swanson to redirect the course made Martin's job untenable, and he left in 1989 to become the executive vice president in charge of research and development at the newly created Du Pont Merck Pharmaceutical Company in Wilmington, Delaware. In 1993, he joined Chiron as Senior Vice President and President of Chiron Therapeutics.

The cloning, in 1982, of TPA, useful in dissolving blood clots in heart-attack patients, was heavily touted to be the blockbuster drug of the decade. Expecting billion-dollar annual sales of TPA upon FDA approval in 1987, Genentech had mobilized a huge force to make and market the drug. With soaring stock prices, with legal battles over patent rights on three continents, and with the legions of vice presidents and staffs to run the business, Genentech was no longer the house of science. Attention was diverted to the outcome of regulatory hearings, to depositions and court trials, and to ways of combatting the claims of competitors offering alternative, cheaper drugs. The scientists were also distracted by concerns about how best to enjoy and invest the bonanza profits pouring in from their ownership of Genentech stock.

A comeuppance was inevitable. Annual sales of TPA were only a third of the projections, the bloated business staff became redundant, and serious scientists were leaving, disaffected by the changed ambiance; plummeting stock prices deepened the gloom. In 1990, 60 percent ownership of the company was acquired by Roche, of Basel, Switzerland, for $2.1 billion, with options to purchase the remainder. Genentech, now a ward of Roche, was freed to concentrate again on science in handsome new laboratory buildings that replaced the old warehouse.

Seeking new horizons in 1991, Dave Goeddel left Genentech for Tularik, which was located only a long stone's throw away from his former labs. This new biotech company was named after the Alaskan river where Goeddel and Professor Robert Tjian, of the Department of Molecular and Cell Biology at the University of California, Berkeley, were fishing when they conceived of this venture. They decided to invite Steven L. McKnight, a member of the Department of Embryology of the Carnegie Institution of Washington, to join them, with plans to explore and mine the complex controls of gene expression. A similar concept inspired Axel Ullrich to start Sugen, in nearby Redwood City, with Joseph Schlessinger of the New York University Medical Center as a cofounder. Goeddel and Ullrich no longer work at the bench, nor do their senior associates in these new ventures. It seems unlikely that Tularik and Sugen will emulate Genentech's story.

Amgen, with Blockbuster Drugs

Amgen, like Genentech, was started on the initiative of a venture capitalist. William K. (Bill) Bowes, Jr., a San Francisco native with a Stanford B.A. in economics (1950) and an M.B.A. from Harvard, had spent 25 years with Blyth and Company, investment bankers in San Francisco. After working largely with emerging electronics companies (such as Ampex, Memorex, and Hewlett-Packard), he was introduced to biotechnology as a representative of his firm on the board of directors of Cetus, starting in 1972. He resigned from the board in 1979, unhappy both with the science and with the management, which was missing milestones and opportunities to exploit genetic engineering.

Bowes had left Blyth in 1978 to start his own venture-capital business (which later became U.S. Venture Partners). He was legally free and clear to listen to Sam Wolstadter, another venture capitalist, who was eager to start something in biotech. Wolstadter had identified Robert (Bob) Schimke, a biochemist and a professor in the Biological Sciences Department at Stanford, as the author of a highly cited paper on recombinant DNA, and he was surprised to find Bob still uncommitted to any biotech venture. Despite a full year of entreaties from Wolstadter, Bob continued to resist the idea of starting a company; but he agreed, at Wolstadter's urging, to meet with people from Tosco, an oil-shale refining company, who were interested in microbe-enhanced oil recovery and were looking for a way (perhaps a microbial way) to remove obnoxious sulfur compounds from crude oil. Although Bob's expertise was with animal cells, he was intrigued with the notion of solving Tosco's problems with genetically engineered microbes. When Bill Bowes was consulted, he neither liked Bob's ideas nor could he persuade Bob to direct a scientific advisory board of a biotech venture with broader objectives. Instead, Bob recommended Winston Salser, a professor in the Molecular Biology Institute at UCLA.

Salser was not only receptive to the idea but ecstatic about the opportunity to organize a biotech company, and he immediately set about assembling a distinguished scientific advisory board consisting of Arnold Berk of UCLA, John Carbon of the University of California, Santa Barbara, Marvin Caruthers of the University of Colorado at Boulder, Martin Cline of UCLA, Norman Davidson of Caltech, David Gibson of the University of Texas at Austin, Leroy Hood of Caltech, Arno Motulsky of the University of Washington, Richard Williams of the International Laboratory for Research on Animal Diseases in Nairobi, Ralph Wolfe of the University of Illinois, and, later, William Rutter of the University of California, San Francisco. Irving Weissman of Stanford University was a consultant, as was Eugene Goldwasser of the University of Chicago, creator of a valuable property: small amounts of purified erythropoietin (EPO), the source of an all-important 29-residue amino acid sequence.

The name of the new enterprise would be Amgen (for *A*pplied *M*olecular *Gen*etics), and the site chosen was the town of Thousand

Oaks, northwest of Los Angeles, near the three academic centers from which half of the board members were drawn. Starting early in 1980, several meetings, over a period of six months, failed to produce any consensus on objectives. In the meantime, Bowes persuaded Wolstadter and other business friends to invest a total of $50,000; a $9,000 share would net $130 million ten years later. The other founders included Moshe Alafi and Franklin Johnson, venture capitalists in the Bay Area, Ed Huddelson, a lawyer, and Ray Baddour, head of the Chemical Engineering Department at MIT, with ties to business and to academia. Alafi had been a founder of Cetus and, like Bowes, had resigned from its board, uneasy with its style and unhappy with its scientific progress. Alafi had also been involved in the start of Biogen, and he had catalyzed Schering-Plough's investment in Biogen and, with that, its entry into biotechnology.

The diverse interests of the science advisors, ranging from chemistry to genetics and from microbes to clinical medicine, were reflected in the many directions Amgen took in the early years: industrial chemistry and microbiology, agricultural strategies, medical diagnostics, and cellular growth factors. In contrast to Genentech's focus on the synthesis and expression of a few genes with clinical importance, Amgen, confused and without adequate resources, desperately needed guidance. Bill Rutter could have provided the leadership, but attempts to persuade him to take the reins were to no avail, both because his roots in the Bay Area were too strong and because his price for control of the operations was too high. The leadership Amgen desperately needed at this juncture came with George Rathmann's taking the helm.

George Rathmann, born and bred in Milwaukee, was attracted to medical research, and he entered Northwestern University as a premed student. At the age of 18, as a sophomore, he applied for admission to its medical school, but he was made so impatient by having his admission delayed for a year that he switched to a career in physical chemistry. When he obtained his Ph.D. from Princeton in 1951, he was greeted by a deluge of industrial job offers.

Rathmann found the Minnesota Mining and Manufacturing (3M) Company, in Minneapolis, most appealing, with its highly creative atmosphere and with the prospect of being able to continue his basic studies of light scattering by polymers. He quickly percolated from research through managerial levels, directing a variety of units that included challenges to Eastman Kodak in photographic films and to General Electric in x rays. Despite innovative ideas, these challenges to the established leaders failed for lack of firm scientific foundations and sustained effort.

In 1973, after 21 years at 3M, Rathmann succumbed to the lure of higher-level management by taking on the presidency of Litton Medical Systems of Litton Industries. After a little more than two years, he realized that poor management, inadequate research and development, and inferior products made for a dismal future. He resigned just as he was about to be fired.

Without a job for several months, he was attracted by Jim Vincent, later CEO of Biogen, to join the Diagnostics Division of Abbott as the vice president in charge of research and development. There, he contributed significantly to bringing diagnostics, which had been an appendage of this established pharmaceutical company, to its current core position with $2 billion in annual sales. Seeking better vaccines and diagnostics for hepatitis B than those prepared directly from the virus, he became aware that an improved synthetic antigen might be created by the new biotechnology. To that end, he was granted a six-month sabbatical leave, which he contemplated spending on a visit to Winston Salser's laboratory, with an introduction from Philip Whitcome, manager of cardiovascular products at Abbott and a former student of Salser's.

During further discussions at UCLA with Salser and others, Rathmann heard the first cries of the newborn Amgen, conceived by venture capitalists and nourished by a board of scientific advisors of diverse talents and interests. At the same time, he looked in on other lusty infants—Genentech and Biogen—and their potential for transforming the world of diagnostics and therapeutics. Despite Abbott's recognition of the success of the diagnostics division under Rathmann, and despite their willingness to have Rathmann start a subsidiary biotech operation, it was clear that a venture as novel as Amgen needed an entirely separate medium. In October 1980, Rathmann accepted Amgen's offer to become its president, CEO, and first employee. His total compensation would be only two-thirds of his Abbott income and would begin only after the first round of financing was completed.

"Everybody's favorite CEO," "the undisputed hero of biotech," and "at the top of everyone's list" are among the accolades accorded Rathmann by biotech industry leaders and analysts. His scientific insights, strong principles, and perseverance were critical to Amgen's success. To these attributes might be added an infectious en-

George Rathmann

thusiasm, a feisty sense of humor, and good judgment in the choice of people, both on the business side and on the science side. He was lucky, too. The first two Amgen drugs, Epogen (EPO, or erythropoietin), for anemia, and Neupogen (G-CSF, or granulocyte-colony-stimulating factor), to combat depressed granulocyte levels in patients undergoing chemotherapy or in patients with blood disorders, are unique and potent agents with minimal side effects. They warrant high prices, and they reap blockbuster profits. But there were times, in the early days, when Amgen's share price had dipped to one percent of current values, when financial analysts made puns about "faded genes" to signify deflated hopes, and immersion in litigation threatened the company's existence.

Amgen would have to go to the financial well six times before it became profitable and secure, only the second fully integrated pharmaceutical company (after Syntex) built from scratch in the post–World War II period. Within four months of joining Amgen, Rathmann completed a private financing of $19 million, by far the largest initial equity financing in biotech history. The principal investors included Tosco and Abbott. The Abbott investment was propelled by Kirk Raab, a group vice president, who left four years later to become the president of Genentech. From its $5 million investment, Abbott derived more than $200 million when its shares were sold after ten years, a sum that would have been twice as great if the shares had been held only one year longer.

With this money in the bank, Amgen's attention would be focused for two years on recruiting staff and doing science, rather than courting new investors and corporate partners. Daniel Vapnek was brought in as the director of research and development. A professor of genetics at the University of Georgia, Dan had been trained in microbial genetics and was enormously impressed by the recombinant-DNA technologies during a sabbatical year (1978–1979) in Herbert Boyer's laboratory at the University of California, San Francisco. Nowell Stebbing, previously the director of biology at Genentech, was engaged to handle many administrative aspects of research and development. Other management positions were filled, notably that of chief financial officer (by Gordon Binder, a Harvard M.B.A. and a Ford alumnus with computer-business experience, who would later succeed Rathmann as CEO).

As expected, Amgen needed another infusion of money, and there was enough enthusiasm, in early 1983, for an initial public offering that raised $43 million at $18 a share (equivalent to $3 in 1993, after three-to-one and two-to-one splits). But, typical of stock prices on the roller coaster of investor confidence, the Amgen stock price slid to half its initial value within three months. By November 1984, the price had sagged further, to $3.75 per share, which is equivalent to 70 cents per share today. These violent fluctuations in the market were no reflection of the steady growth of technological expertise at Amgen, or of the establishment of a subsidiary for advanced oligonucleotide synthesis under Marvin Caruthers in Boulder, or of the wise decision to drop a number of projects and to concentrate on the cytokines EPO and G-CSF. Among Amgen's responses to critical issues, two were especially notable: One was a bold choice to pursue protracted litigation, and the other was a decision to engage a corporate partner in a joint venture. One turned out well, but the other did not.

In 1987, the future looked bright at Amgen. Clinical responses of anemic patients to EPO were excellent, G-CSF was entering early clinical trials, production yields were improving, and a manufacturing plant was about ready for production. Two more rounds of financing had raised another $150 million, and, for the first time, money was in good supply. But in July, the roof caved in. Genetics Institute (GI), the biotech company in Cambridge, Massachusetts, was granted a patent for broad rights to all forms of EPO, including Amgen's recombinant version. Yet Amgen held a patent that covered the manufacture and purification of recombinant EPO. Whereas GI's claim was based on the repurification of urinary EPO and a 1981 conception of cloning an EPO gene, Amgen had actually cloned the gene and had announced it more than a year before the gene was cloned at GI.

In its attempt to reverse the ruling that favored GI, Amgen pleaded, before a federal court, "You can't claim the gene until you actually have it." Suits and countersuits dragged on for four years. Beyond the millions spent on legal fees was the distraction of managers and scientists from important issues vital to the discovery and development of safe and effective drugs. Convinced of the justice and strength of the Amgen position, and aware of the lucrative market at stake, Rathmann, against much seemingly reasonable advice, resisted sharing the license with GI, fought on, and won the war.

The partnership with Johnson & Johnson (J&J) has had a less favorable outcome, and it is still the subject of bitter litigation. To obtain financing at an early, crucial stage, Amgen licensed to J&J the U.S. marketing rights to EPO for uses other than the treatment of anemia in patients undergoing renal dialysis. With widened indications for EPO, the market has grown to unforeseen dimensions and has generated protracted conjecture about rights and royalties. Not uncommonly, a biotech start-up in a joint venture with a large corporate partner finds, in the rare instance of its having developed a successful drug, that its profits are pitifully small.

Amgen, now awash in profits, is poised between operating in the mode of a biotechnology start-up dealing with large cytokine molecules and operating more like a conventional pharmaceutical company with an emphasis on small-molecule, orally active drugs. Perhaps it will evolve, through its several affiliations—a joint venture with Regeneron (see later) and the ownership of a basic-research immunology institute at the University of Toronto—to occupy a middle ground of biotechnology-driven pharmaceuticals.

In 1990, with Amgen successful and well set, Rathmann, at age 62, was eager to turn over the reins and to adopt a less frantic life-style by assuming the chairmanship of Amgen and membership on a few other boards. But, like Alex Zaffaroni, he still had a start-up left in him. ICOS, named for the icosahedral (twenty-sided) shape of some pathogenic viruses, was the answer. It is a venture intended to develop drugs that intervene in cell-to-cell interactions, which are prominent in inflammation. Located near Seattle, ICOS raised $33 million in its initial private financing, shattering the record set at Amgen. Will ICOS and Rathmann succeed? The gestation period for a novel pharmaceutical drug is likely to remain near ten years, so we must wait to see.

Chiron, with Protean Goals

By 1981, William J. (Bill) Rutter had seen many of his colleagues start biotech ventures, and for years he had been itching to launch his own. He wanted to create novel vaccines and to produce scarce hormones with the new technologies. Despite his urging, Abbott, like other major pharmaceutical companies, was not inclined to start a biotech division with the requisite independence and resources. Nor would Rutter ever be able to muster the space, funds, and personnel, even as chairman of the Department of Biochemistry of the University of California, San Francisco (UCSF), to make vaccines and hormones on an industrial scale.

To an extramural enterprise, he could bring scientific and organizational skills honed in academia and experience gained from consultantships with several pharmaceutical companies (Abbott, Eli Lilly, and Merck). With a sense of mission to advance science and its application to medicine, Bill gathered two of his favorite disciples, Edward E. Penhoet and Pablo D. T. Valenzuela, and founded Chiron. Penhoet became the CEO, Valenzuela became the director of research, and Bill, as chairman of the board, shared control of the business and research strategy. Why Chiron? A centaur of Greek mythology, half man and half horse, Chiron was renowned for his healing skills, which he passed on to Apollo's son Asclepius, the Greek god of medicine.

Bill Rutter could have become the CEO of Amgen, with headquarters in Thousand Oaks in southern California, but he wanted to remain in the Bay Area. An invitation from George Rathmann, the newly installed Amgen CEO, to establish an Amgen subsidiary close to Rutter's home, went so far as to identify laboratory space in South San Francisco, and to arrange compensation and stock options for six staff scientists. But anxieties grew, both in northern and in southern California, about which of the two groups would emerge as the dominant Amgen unit. With the founding of an independent Chiron in Emeryville, next to Berkeley, the matter was settled to the relief of all.

William J. Rutter

Bill Rutter, born in 1928, was raised in Malad City, a town of 2700 in a remote recess of Idaho. An early fascination with exotic parasitic diseases (generated by stories told by his paternal grandfather, a British army officer once stationed in India) was leavened with a grounding in hard science and a respect for analytic thinking. After graduation from high school at age 15 and some time spent at Brigham Young University, Rutter felt confined by the parochial atmosphere of Idaho and Utah and accepted a scholarship to Harvard.

As a biochemistry major, exposed to the Harvard greats in biology and chemistry, Rutter was still headed for medical school when he was diverted by a brief research experience back in Utah. Working with Garth Hansen in the Biochemistry Department of the University of Utah Medical School, and with an opportunity to audit some science and medical classes, he opted for science and graduate training in biochemistry. When Hansen moved to the Biochemistry Division of the Chemistry Department of the University of Illinois at Urbana-Champaign, Rutter went with him and earned his Ph.D. in two years. His thesis focused on how closely related sugars—glucose, galactose, and lactose (milk sugar)—are discriminated and assimilated in bacterial and animal cells, with an eye, ultimately, toward explaining the human inborn metabolic error responsible for galactosemia, a disease caused by a severe intolerance to the galactose produced by the digestion of milk.

Wanting to learn more about enzymes and the molecular basis of metabolism, Rutter spent a postdoctoral fellowship year with Henry Lardy at the Enzyme Institute of the University of Wisconsin, and another with Hugo Theorell at the Nobel Institute in Stockholm, before taking a faculty position in his parental department at Illinois. Over the next ten years, he delved

William J. Rutter

ever deeper into enzymatic mechanisms, trained graduate students, taught courses, and climbed to the top of the professorial ladder. But, once again, he felt his horizons confined, lured by genetics and medicine, and driven by the migratory and exploratory instincts of a native son of the Far West.

During a sabbatical year as a Guggenheim Fellow at Stanford University with Clifford Grobstein and Norman Wessels, Rutter probed the biochemical basis of tissue differentiation. These early grapplings led him to studies of RNA polymerase and how the enzyme managed to select certain genes for expression only in an embryo or only in specific tissues, particularly in the pancreas. He saw an opportunity to combine genetic approaches to these aspects of developmental biology in an invitation to join the Biochemistry Department of the University of Washington, where, with Herschel Roman in the Genetics Department as a close neighbor, he could pursue the genetics of simpler, microbial systems. After four years, an even greater challenge was presented. Rutter was offered the chairmanship of the Department of Biochemistry at the University of California, San Francisco (UCSF). There he could link biochemistry and genetics to clinical medicine and try to rejuvenate a moribund department.

Although cited by Ernst and Young, by *Inc.* magazine, and by Merrill Lynch as the "1992 Entrepreneur of the Year" (along with Ed Penhoet, the Chiron CEO), Rutter was even more deserving of such accolades in previous years for his feats of chairmanship at UCSF. Before his stewardship (1969–1982), the department was a collection of scientists of uneven quality, spread over several floors, lacking resources, achievement, cohesion, and leadership, and completely overshadowed by their cousins in Berkeley. For more than a decade, biochemists had been turning down offers of the chairmanship of this nondescript department in a mediocre medical school. Once Rutter was on the scene, he obtained contiguous space, an expanded budget, and enhanced authority. He persuaded the highly gifted Gordon Tomkins to come from the NIH to join him; and, within a few years, they had created a large, topflight unit and had helped to elevate the entire institution to national prominence. After Tomkins' sudden and unexpected death following a surgical intervention, Rutter was able, by skillful recruiting and the adroit use of people and the latent resources of the institution, to create a setting that would also cradle some of the lusty infants of biotechnology, Genentech and Chiron among them.

Chiron, unlike the other biotech start-ups, pursues a wide menu of activities—vaccines, diagnostics, ophthalmic devices, therapeutics, and technologies—all reflective of Rutter's interests and resourcefulness. To sustain discovery efforts in these diverse and competitive areas and to advance them to product development and marketing, Chiron has had to resort to partnering with pharmaceutical companies, initially for modest royalties, but later on in straight fifty-fifty deals.

Based on his chemical training, biological research, and prolonged contact with the problems confronted in the pharmaceutical industry, vaccines appealed to Rutter as the most strategic avenue for clinical application. Prevention creates massive markets, compared with those created by treatment, and a vaccine product is not likely to be threatened by an equivalent generic substitute. Beyond prevention, vaccines may find uses in diagnostics or even in therapeutics. Most exciting at the time were the prospects that genetic engineering might provide the means to obtain vaccines with improved potency and reduced risk and whose efficacy might be enhanced by superior adjuvants. Rutter was already experienced in recombinant-DNA technology (from the cloning and production of insulin, supported by contracts with Eli Lilly), and he felt no hesitation about tackling the creation of a vaccine for hepatitis B to be produced by yeast cells modified to function as vaccine factories.

For a partner in this first Chiron project in 1978, Rutter turned to Roy Vagelos, a friend who had just left the chairmanship of the Biochemistry Department at the Washington University Medical School in St. Louis to become the director of research at Merck, which was marketing a vaccine for hepatitis B prepared from viral particles. Vagelos, who was later to become the CEO of this leading pharmaceutical company, had little faith in the emerging biotech ventures, and he doubted that the large molecules—hormones, interferons, interleukins—that had been targeted as therapeutic agents by these start-ups would ever be successful drugs. Despite Merck's commitment to the hepatitis B vaccine, Merck would rather stick with the small molecules that could be administered by means of the familiar pill or capsule, rather than by injection. Vagelos had no interest in having Merck acquire equity in Chiron, nor was he interested in being associated with any such puny venture. Over the years, he had been offered opportunities to share in the development of erythropoietin and other cytokines and had dismissed them. But he was willing to transfer to Chiron the Merck funding already begun at UCSF for a genetically engineered hepatitis B vaccine.

In the early years, getting the money to operate Chiron was far more difficult than doing the science. Lacking experience in business and in the world of venture capital, Rutter had to absorb many lessons the hard way. In exchange for modest investments, venture capitalists in Boston and San Francisco tried to manipulate the science programs for quick results. Martin-Marietta, a consortium in aerospace and nuclear power, based in Bethesda, Maryland, invested in Chiron with an interest in diversifying in agricultural ap-plications of biotechnology. Like other investors, they kept pressing Chiron to concentrate on their specific goals. The $100,000 that Rutter and Penhoet had invested from their own savings covered Chiron's expenditures for only a week or so. Additional support came from Merck (earmarked for the hepatitis B vaccine), from Ciba-Geigy (to pursue research on the somatomedins [insulin-like growth factors] IGF-1 and IGF-2), and from Novo Nordisk (to ad-vance the production of single-chain insulin in yeast).

This funding carried Chiron along until an initial public stock offering in 1983 netted $20 million and gave the company some breathing space. Three more public offerings and cash from in-vestment bankers expanded that breathing space into running room. The Biocine Company was set up as a joint venture with Ciba-Geigy to replace Merck as the recipient of Chiron advances in vaccine biotechnology; income from selling reagents for blood screening accrues to Ortho-Chiron, another joint venture with the Ortho division of Johnson & Johnson.

Of the many Chiron maneuvers, none match its acquisition of Cetus, one of the earliest and largest of the biotech ventures. Cetus was in financial straits in early 1991 after the FDA had failed to ap-prove interleukin-2 (IL-2) for renal cancer. Sinking under a $150 million debt, Cetus called on the investment firm of Lehman Brothers to search for a buyer or investor. Even though a closed-off skyway physically connected Chiron's headquarters in Emery-ville to Cetus's headquarters, Chiron was not on any list of potential buyers. Inquiries from Chiron about using idle Cetus manufactur-ing facilities, and then a chance parking-lot encounter between the top managers of the two companies, led eventually to Chiron's ac-quisition of Cetus for $880 million. The frisky little fish had swal-lowed the ailing whale whole. By disgorging the Cetus patents for DNA amplification (PCR, or polymerase chain reaction) to Roche, the digestive burden was lessened by $300 million, and an ex-change of stock reduced Chiron's cash outlay to $50 million. But Wall Street was not convinced, and Chiron stock fell.

No one expected, at the time of the sale, that the Cetus product Betaseron (beta interferon) would soon become a drug with enor-mous profit potential. Like alpha interferon, Betaseron had failed repeatedly in clinical trials of its therapeutic efficacy against cancer and viral infections. But attempts to reduce the muscle spasms and paralysis of multiple sclerosis with Betaseron proved remarkably positive in about a third of the patients to whom it was given, and FDA approval of Betaseron was promptly granted in July 1993. To contend with the surge in demand for the drug while supplies of the drug were still inadequate, a computer was used to select 20,000 lucky recipients from a nationwide list of candidates that had been supplied by physicians. When the full number of some 100,000 multiple sclerosis patients in the United States receive treatment, at an annual cost near $10,000 per patient, sales of Betaseron may reach $1 billion. Even given the current emphasis on cost containment, this expensive drug will be a bargain when compared with the cost of hospitalization several times per year.

Chiron has paid down the debt incurred by the Cetus acquisition, and it has managed to get swift FDA approval of Proleukin (interleukin-2) for treatment of very early stage renal carcinoma. Proleukin now has $35 million in annual sales. With enhanced manufacturing capacity and with the Cetus sales forces in place in the United States and Europe, the combined product line of Cetus and Chiron drugs may ultimately justify the stock-price multiple of 100 times earnings per share in 1993.

The adventurous spirit at Chiron comes from the top. David Martin, a former professorial colleague of Rutter's at UCSF, later Director of Research at Genentech, and, finally, an executive vice president in charge of research and development at the Du Pont Merck Pharmaceutical Company before coming to Chiron, contrasts Chiron with other biotech start-ups. According to Martin, the founders of Chiron "are very creative people, and they are not afraid to use their creativity to do things which at first blush people think are crazy. It's risk-taking from the top, which is quite a bit different from the usual biotech company, [where] the people down in the trenches want to take the risk and management [does not want] to tolerate it."

Inevitably, some of the Chiron programs will stumble at one or another of the hurdles along the tortuous path from laboratory to market. The major investment in an AIDS vaccine faces severe competition and discouraging early results. Drugs in the pipeline to treat septic shock, diabetes, and osteoporosis—all complex and heterogeneous diseases—will have to undergo extensive and expensive clinical testing. Yet the future seems bright for the hepatitis B and hepatitis C vaccines, for a new, genetically engineered vaccine for whooping cough, and for diagnostic and ophthalmic products, among others. With five discrete business units—diagnostics, oncology, vaccines, ophthalmics, and technologies—each headed by executives with business backgrounds, and still small enough to be managed by its daring founders, Chiron may justify the current optimism about its future. Confirmation of this optimism was provided, in November 1994, by a Ciba-Geigy investment of $2.1 billion for a 49.9 percent equity interest in a preferred partnership with Chiron.

Regeneron, Pioneer in Neuroscience

More than twenty neuroscience ventures were created in the past decade for the purpose of discovering and developing drugs to treat some of the most ghastly diseases of the nervous system: Alzheimer's, Parkinson's, and Lou Gehrig's (amyotrophic lateral sclerosis, or ALS). Although ambitious to become fully integrated pharmaceutical companies, few or none will make it. Yet the hope for a bonanza in this barren area of medicine is enough to have persuaded investors to risk more than $1 billion on these ventures.

Neurology, by its logical analysis of the signs and symptoms of diseases of the nervous system, is at once the queen of diagnosis in medicine and also the most helpless knave in treatment. Part of this failure stems from ignorance of the complex chemistry of the brain and part from the blood–brain barrier, which, while protecting the highly responsive brain from the chemical vicissitudes of the circulating blood, also prevents access to many drugs that easily reach all other compartments of the body. Thus, the "black box" metaphor, widely used in dealing with mysteries in physics, is especially appropriate and poignant in describing our inadequate knowledge of the functions and aberrations of the brain and of effective treatment of the diseases that afflict it.

Confounding these formidable problems is the mystique that regards mind and behavior as different in substance and function from all other tissues and functions of the body. This notion prevails despite the fact that the molecules and structures in all cells (including neurons) are virtually identical, with an evolutionary heritage shared for billions of years. Accepting that mind is matter, and only matter, is an essential first step in applying the extraordinary technical advances and insights of molecular and cellular biology—so effectively pursued with other body cells and organs—

to the analysis and engineering of the genes and enzymes of the brain.

Notions of an ethereal basis for human behavior should have been abandoned long ago in view of the profound effects of such small molecules as alcohol, nicotine, cannabis, and opiates. Mounting knowledge of the pituitary hormone oxytocin (the pitocin commonly administered during childbirth), only nine amino acids long, is an interesting example. Known for a century to stimulate uterine contraction and to promote lactation, oxytocin has been disclosed by recent research to have startling effects on sexual behavior, inducing cuddling and enhancing the libido in both sexes. Receptors on the surfaces of cells responsive to oxytocin have been found in the brains of men—as well as fish—creatures that have neither uteri nor mammary glands. Clearly, a variety of behavioral patterns may be induced by this very simple molecule.

Another dogma has held that cells of the developing mammalian nervous system are prevented from regenerating in order that intricate neuronal pathways, once established, not be befuddled by the development of fresh synapses. Discovery of the nerve growth factor (NGF), an agent that promotes the development, survival, growth, and functioning of neurons in the sympathetic and sensory nervous systems, encouraged the search for other such neurotrophic factors and still others that likely enhance and suppress their actions. The promise that these natural substances might be effective in the treatment of degenerative diseases, cerebrovascular accidents, and trauma to, and infections of, the nervous system, for which few curative therapies exist, spawned the creation of the many biotech ventures in neuroscience, including Regeneron.

Regeneron Pharmaceuticals, Inc., was founded, in 1988, on the initiative of Leonard S. (Len) Schleifer, who enlisted Eric M. Shooter, Alfred G. Gilman, and other senior scientists with expertise in neurobiology and molecular biology to discover the missing neurotrophins and to exploit their potential as drugs for treating diseases of the nervous system. With excellent academic training in biochemistry and neurology and a wealth of ideas and ambition, Schleifer knew that these difficult discoveries were highly unlikely to emerge either from the pharmaceutical companies, which lacked the vision and talent, or from the academic departments, which lacked the resources. What was needed was a venture in which talent and resources could be joined in one location and with only one goal in mind.

Leonard S. Schleifer

Len Schleifer was born in 1952, grew up in the borough of Queens, and excelled in school, with top SAT scores in math and science. He also found time to win a New York State junior chess title and to engage in team sports. At Cornell, he majored in neurobiology and did research on the purification of acetylcholine receptors of the rat brain and of the stinging electric organ of *Torpedo* (a genus of marine fishes known as electric rays). In 1972, as an applicant to medical school in his junior year, Schleifer was turned down by some of the most prestigious schools, but he was accepted at the University of Virginia.

With his research experience and his academic quickness, Schleifer was recruited to join the new medical scientist training program (MSTP) at Virginia, and to become a candidate for both the M.D. and Ph.D. degrees. His good fortune was to do his graduate work with Alfred G. Gilman, who, at the time, was a young associate professor of pharmacology. Gilman was on the verge of making basic discoveries about the central role of a class of proteins (G proteins) that convey hormone signals from the cell surface to the interior—the work that would earn him a share of the Nobel Prize in Physiology or Medicine in 1994. Schleifer's grounding in biochemistry and his experimental work in protein chemistry would later stand him in good stead at Regeneron, but the medical side of his training had been given short shrift. To correct for his lack of clinical experience, he decided on a residency in neurology, preceded by a grueling internship in internal medicine. He chose the program at Cornell Medical School because, under Fred Plum's direction, neurology was given a dynamic, metabolic orientation, in contrast to the anatomic pathology approach of the Ray Adams tradition at the Massachusetts General Hospital of Harvard Medical School.

When Schleifer completed his residency in 1984, he was a superbly trained neurologist, but by then he was removed from the frontiers of molecular biology: recombinant DNA, cloning, and monoclonal antibodies. After eleven years of intensive postgraduate work, competing for a postdoctoral fellowship had less appeal than accepting an assistant professorship in the Neurology Department at Cornell Medical School, where he promptly received generous research support: the Cornell Scholar's Award, the Physician Scientist Award of the American Heart Association, and an NIH grant.

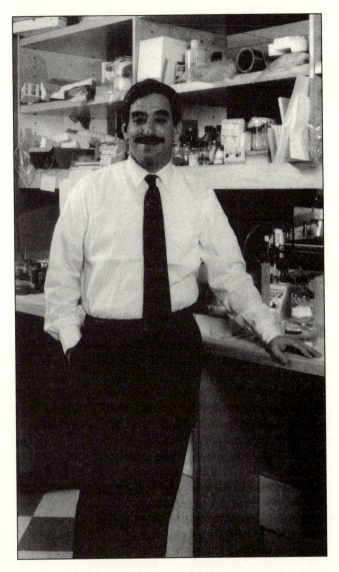

Leonard S. Schleifer

Despite a profusion of funding as an assistant professor of neurology, Schleifer felt that the resources and the academic setting were grossly inadequate to the task he thought most important: the discovery and isolation of neurotrophic factors. The celebrated NGF was the sole trophic factor known in neurology, whereas there seemed to be monthly announcements of new interleukins or cytokines for the immune and hematopoietic systems. Believing that a comparable profusion of factors for the growth and maintenance of neurons must also exist, and that these would be the most exciting drug candidates for treatment of diseases of the nervous system, he resolved to marshal the talent and resources to find them.

Schleifer then sought financing for a company to undertake this project. He enlisted Ira Black, the director of his research group at Cornell, in making a presentation at Syntex arranged by one of the Charles Allen family (investment bankers who had financed the embryonic Syntex in Mexico) who was influential in the company. But John Fried, President of Syntex Research, after this and subsequent meetings, could not be persuaded to make an adequate investment. Nor did Ira Black appear to have any ambitions beyond obtaining long-range funding for his own ongoing research programs at Cornell Medical School. Gilman, Schleifer's trusted mentor and confidante, provided key advice. After many late-night telephone conversations with Schleifer, Gilman recognized that Schleifer's entrepreneurial spirit could not be dampened. Therefore, Gilman suggested that he talk to a respected neurobiologist, such as Eric Shooter, Chairman of the Neurobiology Department at Stanford and the leading investigator of NGF.

Eric M. Shooter

Eric Shooter, born in 1924, had been raised in a small English town north of Birmingham. His father, who had fought throughout World War I, earned a college degree to become a mining inspector. With a scholarship to Cambridge, Shooter majored in chemistry, as had his older brother Ken. Eric's graduate work on peanut proteins, directed by Eric Rideal, a well-known figure in colloid chemistry, emphasized the physical aspects of proteins, a subject he pursued further as a postdoctoral fellow with Jack Williams and Robert Alberty at the University of Wisconsin. Few academic opportunities were available when Shooter returned to England in 1950. After a three-year

Eric M. Shooter

stint at the Brewing Industrial Research Foundation, he spent ten years in the Biochemistry Department at University College, London, where an onerous teaching load and a lack of grants made sustained research impossible. Still, he managed to do comparative studies of fetal and adult hemoglobins, inspired by Linus Pauling's discovery of the genetic basis of molecular structure.

A sabbatical visit to Robert L. (Buzz) Baldwin's lab in the Biochemistry Department at Stanford Medical School was fateful. Besides being given an introduction to DNA, Shooter was recruited by Joshua Lederberg to join the Genetics Department and to assume responsibility for a laboratory, endowed by the John F. Kennedy family, to be devoted to neurobiology. During a return to England (1962–1964) enforced by immigration regulations, Shooter examined protein patterns in the brains of hibernating animals and took courses in neurochemistry. Upon returning to Stanford, he decided to tackle the purification of NGF, which had been discovered a few years earlier at Washington University in St. Louis by Rita Levi-Montalcini (now at the Institute of Neurobiology in Rome) and Stanley Cohen (now at Vanderbilt University). They had shown that NGF-neutralizing antibodies would essentially destroy the sympathetic nervous system of a newborn rat.

Shooter achieved the first isolation of homogeneous NGF, aided by Silvio Varon, who had come from the Levi-Montalcini laboratory in St. Louis, and Junichi Nomura, from the Osamu Hayaishi laboratory in Kyoto. With the sequencing of the protein by Ralph Bradshaw and Ruth Angeletti in 1971 and the demonstration of its effectiveness in promoting axon outgrowth and neuronal maintenance in vivo and in vitro, NGF had come of age and could join insulin and growth hormone as prime candidates for the clinical application of cloning and genetic engineering.

In 1975, Shooter had moved from the Genetics Department to organize a new Department of Neurobiology in the Stanford Medical School, and he served as its chairman for the next twelve years. His command of the NGF field and his broad knowledge of neurochemistry were manifested by insightful lectures and articles. The high regard in which Shooter was held by the neuroscience community led Gilman to urge Len Schleifer to seek his counsel, much as Yoshihiko Nishizawa, of the giant Sumitomo Chemical Company, had done some months earlier. Nishizawa, who was the director of the chemical division (*not* the pharmaceutical division) of Sumitomo, was mindful of the increasing frequency of neurodegenerative diseases in the aging populations of Japan and elsewhere, and he had presciently sought out the leading neuroscience laboratories in the world in order to discuss the possibility of research grants linked to patent rights for their discoveries. In addition to Shooter, Nishizawa identified and met with the other leading investigators of neurotrophic factors: Hans Thoenen and Yves Barde at the Max Planck Institute in Munich, Albert Aguayo at McGill University in Montreal, and Lloyd Greene at Columbia University in New York.

Shooter had been present at Schleifer's Syntex presentation, and he was impressed by him and by his program. He was also reassured by Gilman of Len's intelligence, creativity, integrity, and capacity to achieve an entrepreneurial mission. Shooter agreed to join Schleifer in the new venture. Starting in January 1988, Shooter, Schleifer, and Gilman began assembling scientists who would provide guidance in neurobiology and molecular biology. The board of scientific advisors that met for the first time that spring included Albert Aguayo, Fred Alt, Yves Barde, Michael S. Brown, Moses Chao, Fred Plum, Alfred G. Gilman, Joseph Goldstein, Lloyd Greene, Louis Kunkel, Martin Schwab, Eric Shooter, Hans Thoenen, and me.

I had turned down Shooter's first invitation to join the board, doubting that I could contribute significantly to a venture in neuroscience that was located a cross-country flight away. Shooter persisted and prevailed. Among the considerations for me were a long-standing curiosity about brain biochemistry, a lingering and bitter frustration with my late wife Sylvy's illness (an obscure and unremitting subcortical degenerative disease), and a sense that I might contribute to the creation of a "DNAX of neurobiology," an organization that would do topflight science in a truly academic atmosphere. There were other inducements. I had remarried, and there was the pleasant prospect of two weekends a year in New York with my wife Char in the company of old friends already committed to the board and others whose acquaintanceship would be attractive. The equity that Regeneron offered its board members was very generous, which promoted loyalty and active participation.

When Sumitomo finally decided to make concrete offers to Shooter, Thoenen, Barde, Aguayo, and Greene, they learned that these heavy hitters had already signed on with Regeneron, Schleifer's first coup. Merrill Lynch, with an investment of $1 million, put Regeneron in motion and placed its own George L. Sing, then 34 years old, on the board of directors, which initially also included Charles A. Baker (CEO of The Liposome Company, Inc.), James W. Fordyce of Prince Ventures, and Fred A. Middleton of Sanderling Ventures, in addition to Schleifer, Shooter, and Gilman. (Sing's unusual devotion to the science at Regeneron has kept him on the board even after he left Merrill Lynch to start his own venture-capital firm.) Schleifer rented a one-bedroom apartment in a Cornell Medical School housing unit and started managing Regeneron on a one-day-a-week basis. That schedule lasted only one week before he resigned his academic post to devote full time to the business. Within nine months, Schleifer found four investors who ventured a total of $6 million to enable him to build laboratories and hire staff.

Regeneron, in Tarrytown, N.Y.

Schleifer was qualified to direct the research, but Regeneron clearly needed a CEO to manage the business. Or so it seemed, until Schleifer performed these operations with a zest and mastery that more than filled the bill. Concerns that he might lack organizational talent, or that he might be consumed by sharks in financial and pharmaceutical deals, were repeatedly dispelled. With increasing confidence in his own judgment and with unwavering support from his boards, scientists, and managers, Schleifer staffed the expanding laboratories, directed the research, arranged financing (including an initial public stock offering, which netted $92 million), and later on managed the design and construction of a pilot plant and drug-production facility. With all that, he still found time to fulfill a compassionate concern for patients by serving as an attending physician in neurology at the New York Hospital of the Cornell Medical School.

Most neuroscience ventures focus on one or more of the common, intractable diseases of the nervous system. There was considerable pressure on Regeneron from several sources to follow this obvious course. On the other hand, a number of us among the scientific advisors supported Schleifer's inclination to focus on molecular biology in order to discover new neurotrophic factors without regard for their immediate application. Thoenen and Barde in Munich were in the vanguard of such efforts, and they had purified and cloned a rat brain-derived neurotrophic factor (BDNF) that turned out to be the second member of the NGF family, a discovery they would share with Regeneron for cloning the human gene and protein. It could be anticipated that, as with the interleukins, a family of neurotrophins would be found with specificities for various neurons functioning at different stages of growth, differentiation, and maintenance. With these factors in hand and patented, Regeneron could distribute them widely and thereby enlist the entire academic neuroscience community in revealing their physiological functions and clinical potential.

Beginning in March 1989, the Regeneron laboratories were located in space previously occupied by Union Carbide on a 275-acre campus near Tarrytown, New York, in the lovely hills of the Hudson River valley. Designed by Kornberg Associates, the labs were colorful and functional in the style of those at DNAX in Palo Alto. Only a forty-minute drive from Manhattan and close to affordable suburban housing, the setting was ideal for bringing scientists to New York, a region with rich academic and clinical resources, yet biotechnology-poor when compared with Boston and the San Francisco Bay Area.

For direction of neurobiology, Regeneron turned to Ronald M. Lindsay, age 41, who had been head of cell biology (focusing on neurobiology) at the Sandoz-supported laboratories at University College, London, for four years and had extensive experience in neuropharmacology and neurophysiology. Mark E. Furth, age 37, came to head up the molecular-biology division. After extensive training in microbial genetics, Mark had spent about ten years in studies of molecular oncogenesis, first at the National Cancer Institute, then at the Memorial Sloan-Kettering Cancer Center, and most recently at Oncogene Science, Inc., with responsibility for their program in cancer diagnostics.

To finance research and the subsequent development of products, joint ventures were sought with several of the major pharmaceutical companies. Milestones of progress over a five-year span were required and reasonable, but the conventional small royalty of 6 percent to 10 percent for Regeneron was not. Schleifer believed that an even split, with an equitable sharing of development costs, was the only fair arrangement—which, in most cases, was a hard sell. Another difficulty with a joint venture that might likely arise later was the possibility that a discovery at Regeneron would have to wait in line behind the discoveries made within the other company to obtain the considerable resources needed for its development.

Monsanto-Searle was approached in a meeting held in Dallas in 1988. Before the presentations were made by each side, I thought it presumptuous of Regeneron, still in swaddling clothes, to be there. How could it offer something of value to Searle, with its established neuroscience program coupled to the vaunted chemical strengths at Monsanto? At the end of the day, I had the answer. It was clear to me that a poorly focused effort within a huge organization would go nowhere, but that Regeneron, with fire in its belly and Schleifer to stoke it, would run a strong race. Typically, the large companies did not see it that way. Monsanto-Searle showed no interest, but, fortunately for Regeneron, Sumitomo came through with a $10 million investment with few strings attached, an infusion of cash and moral support at a crucial time.

Shooter's friendships at Ciba-Geigy in Basel had gained Schleifer an audience for joint-venture proposals. After year-long discussions, a deal appeared settled for $40 million. When negotiating about the minutiae of a host of legal issues was over, Regeneron was offered only $33 million, a still smaller share of the profits, and some lame excuses for the discrepancies with the agreed-upon initial proposal. Schleifer simply walked out. Less than a year later, he concluded a far more generous arrangement with Amgen: $53 million with equal profit sharing.

In 1991, Regeneron went public with a stock offering managed by Merrill Lynch. Enthusiasm at the preceding presentations was great enough to encourage Merrill Lynch to sell more shares than originally intended, 4.5 million instead of 3 million, at a price of $22 per share. Although shares had been allotted to presumably conservative investors, nearly half turned over the very first day. So much for long-term investment in biotechnology. Critical comments by analysts about the price and number of the shares in the offering contributed to a decline in Regeneron market value to a low of $7.75 per share one year later, with a gradual recovery to a high of $21.50 per share in July 1993. Then, in June 1994, with the failure of an extensive and expensive clinical trial of the human ciliary neurotrophic factor (CNTF) to treat ALS, the stock plummetted to a low of $4 per share.

Regeneron's dedication to exemplary science as the strategy for corporate success is strongly supported by the nine members of its board of directors, which includes four distinguished scientists: Brown, Gilman, Goldstein, and Shooter. All felt that the most important tactic for obtaining excellent science is obtaining excellent scientists. This Schleifer achieved by an all-out effort to recruit George Yancopoulos, a "franchise player" with abilities and impact comparable to those of Dave Goeddel at Genentech and Kenichi Arai at DNAX.

George Yancopoulos

George Yancopoulos, born in 1959 and raised in the Astoria section of the borough of Queens, the largest Greek colony outside of Athens, knew no English when he started school and was thought to be rather slow in learning. He then proceeded to excel at every stage—valedictorian at the elite Bronx High School of Science, winner of a Westinghouse scholarship to Columbia, and valedictorian at Columbia. An undergraduate interest in structural chemistry, through research with Jonathan Greer, became directed to molecular biology when he encountered a paper by three Greeks at Harvard—Argiris Efstratiadis, Tom Maniatis, and Fotos Kafatos—on the cloning of recombinant DNA. Diverse interests in science and medicine, and parental influence to pursue a secure professional career, led Yancopoulos to an M.D.-Ph.D. program at Columbia College of Physicians and Surgeons.

For a research advisor, Yancopoulos eventually chose Fred Alt, who was about to join the faculty from a postdoctoral fellowship with David Baltimore at MIT. Interest in immunology and a fine rapport with Alt, an ambitious young scientist (who did his graduate work with Robert Schimke at Stanford), developed into a collaboration that revealed key elements of the genetic recombination that creates the huge variety of antibodies. Upon completing the curriculum, he found there were many pressures to cement his medical studies with clinical training, but Yancopoulos clearly preferred science. He would have been happy to continue the congenial and productive association with Alt indefinitely, but convention demanded that he strike out on his own.

Awarded a prestigious Markey Fellowship, Yancopoulos was besieged by offers and inquiries to take this grant to start an

George Yancopoulos

assistant professorship at Columbia, UCLA, Einstein, MIT (at the Whitehead Institute), or Stanford. He chose Columbia with the promise of new space and a fast, tenure-track future. Then a call came from Len Schleifer, inviting him to dinner and to consider a job at Regeneron. There was no chance he would take it seriously, but "a few free steaks" might be worth the time to listen. He did not expect the barrage of arguments and inducements: a chance to explore a new frontier in neurobiology equipped with the powerful technology he had acquired in work on immunology. There was also a rekindling of old interests in limb regeneration, and an ambience of scholarship, communal sharing, unlimited resources, and freedom from bureaucratic annoyances.

Schleifer's relentless persuasion was backed up by Ronald Lindsay, Mark Furth, and members of the scientific advisory board. (I told Yancopoulos about the meteoric career of Kenichi

Arai at DNAX, describing his pioneering role in molecular immunology, which had exceeded what would have been possible in a university setting and led, in only a few years, to numerous offers of professorships.) Naive and overwhelmed, seemingly otherworldly but strongly influenced by his father (who attended the interviews with Schleifer), Yancopoulos was aware that the financial rewards at Regeneron would benefit his family in ways that the impecunious recompense in academia never could.

Yancopoulos's decision to join Regeneron was greeted with disbelief by the illustrious professors who had invited him to their faculties. "Don't throw away your career," was the common refrain. What Yancopoulos gained quickly at Regeneron was an education in neurobiology and fertile ground for his seeding it with molecular biology. Cloning of novel neurotrophic factors, exploration of their sites of action, and discovery of their receptors went into high gear with ample equipment, technical assistance, and acquisition of young scientific talent. In 1992 and early 1993, the sweep of basic research, driven by Yancopoulos, led to twenty papers published in the most celebrated journals, which established Regeneron as a leading neuroscience institute much sought after for postdoctoral fellowship training.

To expand its research effort and to develop its most promising discoveries, Regeneron, like other biotech ventures, faces a most daunting dilemma. The costs of manufacturing facilities and clinical trials are staggering. To go it alone requires additional rounds of financing, and it diverts attention from the research enterprise. Yet, to sell a discovery to an established pharmaceutical company forfeits control of the development process and circumscribes the future of the smaller venture.

Regeneron gambled heavily on developing CNTF, one of the most promising of the neurotrophic factors, first purified and cloned from the rat by Thoenen and his group and shown, in animal models, to be effective against diseases of peripheral motor neurons. With the prospect of controlling ALS with CNTF, a potentially life-saving drug with a large market, Schleifer installed manufacturing plants, organized clinical testing on more than 700 patients, developed regulatory divisions, confronted patent claims by Synergen, and secured additional financing in a depressed market. Then, after nine months of the clinical trials, the results were judged a failure. At the doses used, the toxic side effects offset any beneficial actions the drug may have had.

With regard to other Regeneron projects, BDNF and certain other factors in the NGF family, development will be shared with Amgen, but the costs will also be mammoth. The stakes are high, but faith persists that stellar science with adroit management is the best recipe for success. Clearly, this faith is shared by Roy Vagelos, who assumed the chairmanship of the Regeneron board of directors in January 1995. Having retired from Merck after a decade as its CEO, Vagelos was attracted to Regeneron by its "combination of top-caliber people, leading-edge science, and long-term commercial potential."

CHAPTER 8

Pros and Cons of Biotech Ventures

Effective use of new knowledge depends on its rapid communication and the zeal and skill of its entrepreneurial developer. Free and open exchange in laboratories, academic and industrial, should be followed by prompt and full disclosures in meetings and publications. Secrecy bred from fear of loss of credit and recognition is a human trait often fostered both in academia and in industry. The intrusion of investment and profit, basic features of the business world, can intensify any inclination to guard and withhold information. The intent of the patent system is to reward the inventor and then, by disclosure, to promote the widest use of the invention. Yet, legal complications and wrangling have corrupted this basic purpose.

Secrecy and Patents

Secrecy is corrosive; it makes even less sense in industry than in academia. In a competitive academic field, the disclosure of a reagent or a procedure, or the hint of success in some direction, may lead another scientist to reproduce and publish a result quickly to gain priority for an important discovery. By contrast, discoveries in the industrial world far exceed the resources needed to pursue them. What matters most is making a shrewd choice of which discovery to develop, because each of the costly and time-consuming hurdles of clinical testing, regulatory approval, quality control, and marketing is crucial in the success of a product. In the research-and-development expenditures of a pharmaceutical company,

231

more than 90 percent is spent on development, whereas, in academia, discovery with little follow-up can establish the discoverer's reputation. Inasmuch as the bulk of discoveries at the frontiers of all of the medical sciences are made in academia, being open and well-informed scientifically is the best insurance for a company that makes a $300 million, ten-year investment in the development of a new drug.

When I express these convictions about the open atmosphere at DNAX and its value to Schering-Plough, listeners have been polite but almost invariably skeptical. Some may ascribe my views to naiveté about the demands of a competitive industrial world, but none should ignore the experience of Alex Zaffaroni. His innovative postdoctoral programs in steroid research at Syntex for more than three decades, which welcomed gifted scientists from competing companies as well as from academia, repeatedly proved to be of great benefit to Syntex. At a forefront of research, the preponderance of new knowledge comes from sharing information with others, which is impossible in an atmosphere of secrecy.

The most important ingredient in any scientific enterprise remains the quality of the scientists. To attract and retain creative and productive scientists, industrial management must provide an open atmosphere that encourages free discussion and objective analysis of ideas and reports of progress (as well as reports of failure), both from within the organization and from the outside world. The best job tenure for scientists is recognition for their achievements, promptly and forthrightly published and communicated at professional meetings. In such a setting, which is conducive to maintaining a flow of students, postdoctoral fellows, and visiting professors, scientists can enjoy the confidence that their creative talents and energies will have their fullest expression and recognition.

The constitutional purpose of the patent system of the United States is to "promote the progress of science and the useful arts." This is achieved by encouraging investment in high-risk research; early patent disclosure encourages others to innovate, improve, and design around the patented invention. The intellectual property, acquired at great expense, is protected for commercial development by excluding others from exploiting the claimed invention for a period of time deemed sufficient for the inventor to profit from it. Without patents, inventors would often be motivated to preserve their inventions as "trade secrets."

Intense competition (as currently found in biotechnology), a scarcity of funding, and personal motives all strain and distort the patent system. The convergence on a few "hot areas"—the cytokines of the immune and hematopoietic systems, the neurotrophins of the nervous system, the cell-surface receptors for these cytokines and their antagonists—has generated an avalanche of patent applications and issued patents, and it has also generated further attempts to contest and circumvent these patents. Academic institutions (whose scientists planted the seeds of biotechnology and trained the manpower for its growth) are jealous, especially in financially stringent times, of the fortunes reaped, as a result of their efforts, by entrepreneurs, investors, and business-oriented faculty. Patent litigation in industry and academia and the new patent consciousness in academic circles have become serious and highly counterproductive preoccupations.

The standards for granting a patent by the U.S. Patent and Trademark Office have oscillated widely since Thomas Jefferson started the agency two centuries ago. At one extreme, inventors were encouraged to make improvements, as Thomas Edison did to perfect the light bulb. At the other extreme, inventors were discouraged, as with the current practice of allowing very broad claims for numerous patents in biotechnology. A recent example is the patent granted for a heat-stable DNA polymerase, an enzyme previously well known in the literature, that excludes others from virtually all of its uses, as well as from some uses of a whole range of related enzymes.

In the late 1960s and early 1970s, most patents that were subjected to litigation were held to be invalid. In 1982, with the formation of the Court of Appeals of the Federal Circuit, which was specifically designed to determine appeals concerning intellectual property, the situation was reversed and the enforceability of patents was strengthened. Although patent litigation may prove to be expensive, patents are often a small company's most valuable, tangible assets, allowing them to compete effectively with much larger firms.

Still, the validity of most biotechnology patents—for insulin, TPA, erythropoietin, GM-CSF, alpha interferon, and CNTF—have been contested on the grounds that they lack novelty and are obvious extensions of prior discoveries. In each of these cases, some of them still pending, the scales of justice could be tipped either way. Most worrisome is that many patents have been granted with very broad claims that may encompass "prior art" (knowledge in the public domain) or cover unforeseen future developments and thus impede access to important technology and its wide application.

Some of the fault rests with a Patent Office staff that is insufficient to cope with the enormous increase in workload, that is inadequately trained to judge the science in new and highly specialized areas, and that lacks modern equipment for data storage and retrieval. Assigned a large number of complex patents to judge in a very few hours, examiners often quit, after a year or two, for more relaxed and lucrative jobs in industry. No wonder that an underpaid, overworked, and beleaguered examiner might grant a broad patent with a sigh, "Let them fight it out in court." And so they do.

The protracted litigation that has enveloped many biotechnology products and processes has been debilitating and demoralizing. Costly and lengthy legal proceedings drain the limited financial resources of a small company and, even worse, divert them from their central missions of research and development. The time and attention of the scientific staff, as well as of the managers, become focused on legal maneuvers, depositions, and court proceedings. The quest for knowledge, innovations, and concerns for human welfare are the victims, because they become secondary to the very survival of the whole venture, which depends on the legal outcome.

With all its faults, the patent system is still the best means to assure that discoveries will be rewarded. Without patent protection, there would be insufficient incentive for the enormous amounts of private investment required in biotech ventures. Yet, the unreasonable breadth of many patent claims, the appropriation of knowledge from the public domain, and the inadequacy of our legal system to adjudicate the validity of patents and the priority of inventions has led to diversion of resources, injustice, and frustration.

A major problem is in making a scientifically unsophisticated judge or jury understand the complex scientific issues in a biotechnology patent case. Issued patents are presumed to be valid. Although this presumption can be challenged, even clear and convincing evidence can be obfuscated. As will be evident later in this chapter from the litigation in which Du Pont challenged the Cetus PCR patent, the difficulty facing the judge and jury in sifting the complex scientific evidence presented the patent challenger with the far heavier burden. At the conclusion of a trial decided by an utterly confused and scientifically untrained judge and jury, winners proclaim that "the system works," while, on another occasion, as losers, they bemoan "the miscarriage of justice."

The capricious outcomes of numerous patent litigations demonstrate that dependence on the judicial system is, and will remain, precarious. It would be far better to improve the patent process. Increasing the number and quality of examiners, decreasing their workload, and equipping them with up-to-date software designed specifically to simplify their job, would be a good start. A more dramatic improvement might come with the use of ad hoc committees of scientific advisors, like those convened by the FDA, to determine the validity and scope of the patent claims.

At the stage of a patent reexamination, requested either by the patent holder or by a challenger, such a peer-review panel, selected by the Patent Office for their expertise in the subject, might be asked to consider the merits of the case, with testimony presented by both parties. Following the reexamination, should the opinion of the examiner be challenged in court, the judge and jury would then be armed with the expert panel's advice. They could then be guided by the conclusions of an impartial, knowledgeable panel, rather than by the testimony of experts selected and paid by the litigants. The judgments rendered would then be based on the substantial arguments of truly expert scientists rather than on the cursory and biased opinions of so-called expert witnesses. Not only would the judgments be improved, but the quality of patent applications would also be enhanced, and enormously expensive and distracting litigation would be minimized.

In the current scene, a small biotechnology company is well advised to engage an in-house patent attorney with training in, as well as knowledge of, the science. As a participating member of the research teams, conferences, and impromptu seminars, the resident attorney becomes indispensable. By catalyzing interactions between research groups, such an attorney can expedite and improve manuscripts rather than retard them, can recognize discoveries that are appropriate for applications, can suggest extensions of promising findings, and can reassure investors that hard-earned discoveries are protected and not squandered.

The independent legal department of a large company or a private patent-law firm can supply a range of talents, technical expertise, and experience, as well as the impartiality and objectivity that cannot be provided by an in-house attorney. There are occasions when a law firm will need to be consulted by the resident attorney. However, the large departments and firms, with their established attitudes and schedules, accommodate less readily than the resident attorney the individual research styles and lifestyles of the motley collection of scientists in a biotech venture. Commanding a scientist's attention to legal matters is easier when it is initiated by a sympathetic and conscientious in-house attorney.

The Cetus Patent for PCR

In 1985, a patent for PCR, the DNA polymerase chain reaction (used in the amplification of small samples of DNA), was granted to Kary B. Mullis and assigned to the Cetus Corporation. This truly remarkable technique enables a tiny fragment of DNA to be amplified as much as a billionfold. DNA, whether obtained from preserved traces of a fossilized prehistoric creature, or present in a single virus particle among a million cells, or found in a bloodstain or saliva at the scene of a crime, can be detected and characterized with extreme sensitivity. Nearly certain determination of paternity is made possible by DNA amplification, as are a thousand other uses in basic studies of evolution, inherited disease, infection, mutagenesis, and carcinogenesis. For immediate application of this technique to the diagnosis of disease, a license was granted by Cetus to Hoffmann–La Roche (Roche), virtually excluding Du Pont and other competitors in the diagnostics business from access to this powerful technology.

In 1990, I was called by a lawyer from Du Pont, and a Washington, D.C., patent-law firm retained by Du Pont, for advice regarding findings in papers published in 1969 by H. Gobind Khorana and his associates, then at the University of Wisconsin. In pioneering studies of the chemical synthesis of DNA, they had used DNA polymerase to amplify the minute quantities of the stretches of DNA that had been laboriously prepared by synthesis. I must have read these papers when they had appeared twenty years earlier, and it is likely that I found them unremarkable because they applied the well-known properties of DNA polymerase, which my laboratory had discovered over the preceding decade. The capacity of the enzyme to copy DNA (in the form of a primed template) in a variety of contexts had been referred to in our publications over and over again. I informed the Du Pont lawyers that the 1969 Khorana papers had described the essence of the PCR technique and that I was willing to testify to that effect as an expert witness in the suit to challenge the Cetus patent.

Before the trial began in the U.S. District Court in San Francisco in January of 1991, Roche had petitioned the Patent Office to reexamine the 1985 Cetus patent and had obtained its reaffirmation. Unfortunately for Du Pont, there had been no opportunity then to discuss fully the Khorana papers as evidence for the prior discovery of PCR, nor did the examiner really understand the basics of the science in question. Neither the jury of six nor the judge had any knowledge of biochemistry, let alone the background to understand this special area of DNA enzymology. Such was the setting in which the case, with a billion-dollar market at stake, was argued for almost two months. The testimony filled a 21-volume court transcript; the suit cost each of the litigants many millions of dollars.

Never having been in a courtroom and with no experience in litigation, I found participation in depositions and preparations for a jury trial fascinating and challenging. The Du Pont lawyers were dedicated and scrupulous with the facts and in their adherence to legal procedure. The comraderie of engagement in a just cause was heightened by the uphill battle to be fought against entrenched adversaries. The Patent Office had ruled twice in Cetus's favor, the local press framed the suit as an assault by Du Pont, the eastern giant, on a puny but valiant enterprise across the bay in Emeryville, and there was little chance that the judge and jury could analyze the arcane facts required to reach a proper conclusion. Yet there was hope that justice would prevail.

"Why," the Cetus lawyers asked, "had the PCR technique languished, if it had been known for fifteen years?" Our response was that the accessory technology essential for the effective use of PCR had not been available. Techniques for determining the sequences of DNA and for synthesizing large quantities of primer DNA were not introduced until a decade after the Khorana reports; in the meantime, the cloning of recombinant DNA provided a facile means of amplifying large stretches of genetic material. To the beguiling recitation by Kary Mullis of how the idea of using DNA polymerase came to him while on a drive with his girlfriend, Du Pont countered that it was indeed an inspired packaging of a good idea and one deserving of commercial advantage, but, as with the rediscovery of the wheel, it was not a patentable invention.

It was the judgment of the Nobel Committee for Chemistry that Kary Mullis be awarded a share of the Prize for 1993 for the PCR technique. Alfred Nobel's will directs that the prize be given for a major discovery. Unlike election to the National Baseball Hall of Fame, the awarding of the Nobel Prize is not meant to honor a lifetime record of achievement. Not only has this criterion for selection eliminated many deserving scientists, it has also made laureates of a few who were caught in the spotlight of a timely discovery but who did little other work of distinction before or after the work for which the award was made. Some of the glamour unique to the Nobel Prize may, in fact, attach to the chance that it affords any scientist, in one inspired experiment, to become, as in a lottery, a major winner and celebrity. For an off-the-wall encore of PCR technology, Mullis has founded Stargene, a company to make and sell replica DNA fragments from the skin and hair of pop stars to satisfy a presumed public appetite for a piece of their heroes and heroines.

Among the many obfuscations by the Cetus lawyers in the Du Pont challenge was the statement that the DNA used by the Khorana group would have assumed a hairpinlike conformation and thus would have been unsuitable for PCR amplification. Another contention was that the language and format of the descriptions of DNA amplification in the Khorana reports were vague and inconclusive, an argument disingenuously supported by the expert witnesses for Cetus.

Both arguments were rebutted. Bruce Wallace, an expert on DNA structure and behavior, went back to his laboratory at the City of Hope National Medical Center in Duarte, California, and repeated the Khorana experiments with DNA samples identical to those used in 1969, available now in plentiful quantities, and obtained excellent PCR amplification.

As for a clearer exposition of the Khorana findings, Ruth Kleppe, the widow of Kjell Kleppe (a postdoctoral fellow from Norway and the first author of one of the 1969 reports) and herself a coauthor, found just such an account in the draft of a presentation that Kjell had made at a Gordon Conference in New England that year. Workers clearing the old biochemistry building at the University of Bergen had come upon some of her late husband's papers and had stacked them in her office. Among them was the detailed report he delivered at the Gordon Conference. Although his notes were not admissible as evidence, and although the proceed-

Kary B. Mullis. (Photo courtesy of Scan Press, London.)

ings of these conferences are not published, we did have a list of the scientists who had attended that conference. Among them was Stuart Linn, Professor of Biochemistry at the University of California, Berkeley. When I called Stu to ask whether he recalled the Kleppe presentation, he found his notes of that conference and instantly said, "Hey, this is PCR! I have been using Kleppe's diagrams to teach PCR to my biochemistry classes for years. In fact, Kary Mullis was a graduate student in one of those classes."

The Du Pont and Cetus lawyers both took depositions from Linn over that weekend. When the Du Pont lawyers asked that Linn and Wallace be called as witnesses, Judge Marilyn Patel would not admit their testimony. As recorded in the official transcript, she said:

> The whole notion about preparation for trial is that the experimentation doesn't go on during the trial if you come up with yet another theory or evidence.
>
> I mean, that, that may be the province of scientists. I mean, any notion that we're here searching for the ultimate truth, you and I, you know, we all know is nonsense.
>
> The question is, who at the time the case is set for trial has the burden of proof, meets that burden of proof at that time, and that's it.

The Du Pont lawyers argued that the Linn evidence was discovered by accident during the trial and that the Wallace evidence was obtained to rebut speculations of Cetus witnesses; these speculations were first disclosed by Cetus during the trial and could not have been known beforehand. Even so, the judge rejected these arguments saying, "Du Pont has, as you know, ample resources to have discovered this in advance if it really puts its mind to it."

It is possible that, had Linn and Wallace been allowed to testify, the jury might still have found for Cetus. Du Pont had to overcome the authority of the beribboned, august Patent Office, the confusion created by the Cetus lawyers and their expert witnesses, the homey image of the off-beat inventor portrayed by Kary Mullis, and the mismatch between an apparent underdog, Cetus as David, pitted against Du Pont, a rich and powerful Goliath.

After the verdict, Chiron purchased Cetus and sold the PCR patent portfolio to Roche for $300 million. Roche has aggressively protected its investment by restricting the use of DNA polymerases and their functions, technology that had been in the public domain for more than twenty years. Licenses are generally denied by Roche to industrial applicants who seek them. Companies worldwide, denied access to the PCR technology by Roche, spend considerable human and financial resources to circumvent the patent, instead of advancing the technique and widening its application. An academic scientist who wished to prepare and use his own thermostable DNA polymerases for PCR was threatened with a lawsuit. In this instance, the scientist's university was unwilling to undertake the costly defense against a suit, and so it acquiesced to buying the expensive reagents from Roche.

The wondrous PCR technique, for all the benefits it provides for the progress of biological science and medicine, is tarnished by inappropriate commercial exploitation of knowledge that was paid for, and belonged to, the public. Innovation and the broadest possible use of DNA polymerases have been stifled. Those of us who actually discovered the DNA polymerases and who described the features basic to PCR are now restricted in the source of these enzymes and in their uses in industrial applications.

Stanford–UC Patent for Recombinant DNA

The patent filing by Stanley Cohen of Stanford University and Herbert Boyer of UC San Francisco in 1974 for the use of recombinant DNA to clone genes in bacterial, plant, and animal cells stands in marked contrast to the Cetus PCR patent. Issuance of the recombinant DNA patent (assigned formally to Stanford University) by the Patent Office in 1980 acknowledged that the claims, by Cohen and Boyer, for the use of plasmids and restriction nucleases constituted a novel invention of recombinant-DNA technology.

Yet, this feat was only a demonstration of principles and approaches established by many others. Although the Cohen–Boyer patent was even broader in its claims than the PCR patent, the annual license fee of $10,000 for access to the Cohen–Boyer patent appeared modest. The cost was a genuine burden for many small biotech start-ups, but it was still tolerable enough to discourage users from mounting an expensive legal challenge. With the sprouting of a vast number of biotech ventures, the financial returns have been handsome. The patent income, shared equally by Stanford and the University of California (UC), is expected to gross

$170 million by the time the patent lapses in 1997, the largest sources of such income in the history of universities.

Stanford's decision to patent the recombinant DNA technique and the breadth of the claims upheld by the Patent Office were widely noted, and they set patterns, both in academia and in industry. "If Stanford can do it," came the common refrain, "why can't we?"

Despite the financial benefits for research and training at Stanford and UC, was it appropriate to patent the recombinant DNA technique? Was it right to arrogate knowledge and reagents acquired worldwide by a generation of scientists supported by public funds to pursue basic research? Would this departure from precedent in the biological sciences prove beneficial or detrimental to the acquisition of knowledge for its own sake in academia? Would it help or deter more fundamental studies in industry?

These questions were initially troubling to Stan Cohen. They also troubled Paul Berg, who had taken a prominent role in formulating guidelines for the proper conduct of DNA research. Were a patent sought, Berg urged that any income be administered by a trust for broad use in training and research. Nevertheless, the university administration filed the patent applications for Stanford and UC; with financial success, the academic community has come to accept the propriety of patents and their aggressive pursuit. The Stanford Office of Technology Licensing takes a 15 percent fee off the top of the patent income, deducts expenses, and divides the remainder equally between UC and Stanford. The Stanford share is divided three ways: the inventor, Stan Cohen, gets one-third; the inventor's departments, Genetics and Medicine, share one-third; and the Medical School, in which these departments are located, gets the other one-third. But the Department of Biochemistry, and the many people in it who laid the groundwork for recombinant DNA, and who actually helped to do the experiments on which the patent is based, have, for their squeamishness about patents and commercialization, been largely uncredited for the invention and unrewarded for their efforts.

In these financially troubled times, scientists and institutions dream of plums and scramble for crumbs. A lucrative patent, such as the Cohen–Boyer patent, is a plum dream come true. Yet too little account has been taken of the cost of the prize and the bargains made to acquire it. The mind-set of students and faculty necessary to perceive the patent potential of an invention and to expend the time and energy to pursue it comes at the expense of probing its underlying scientific basis and exploring a variety of its extensions with no immediate practical objectives. Even so, in the atmosphere of universities today, failure to pursue a patent opportunity may be regarded both as irresponsible and as the greater loss.

The Stanford Licensing Office, with a staff of twenty and an annual budget near $2 million, is concerned as it approaches the "cliffs," the years when the most lucrative patents expire. The carrot to entice university scientists and engineers to "think patents" is the personal share awarded to the inventor and the approbation of his or her department, school, and university. The natural reluctance to dilute one's share of an award and the legal need to identify the "correct" inventors may reinforce reclusive and secretive tendencies and may discourage making communal efforts for discoveries.

Entrepreneurial university departments and institutes have entered into more than a thousand partnerships with companies. To name just a prominent few: Harvard University with Hoechst-Roussel and Du Pont, Washington University with Monsanto, and a widely criticized agreement of the Scripps Research Institute in La Jolla with Sandoz Pharmaceuticals. In return for financial support of a significant part of the research budget, the company is generally granted privileged access to the scientific staff and their research progress and the right of first consideration for patents and licenses. Inevitably, the company also profits from the research funded by federal and other agencies and from contact with related departments in the academic institution. Whether stipulated or not, free communications and exchanges with academic colleagues, and especially with scientists in other companies, may be impeded or discouraged.

In an agreement announced in 1992 between the Scripps Research Institute, the largest, independent, nonprofit, biomedical research organization in the United States, and the U.S. subsidiary of the Swiss Sandoz Corporation, Scripps was to have received $300 million over ten years in exchange for Sandoz being given rights to patents for nearly all of Scripps's discoveries. When this contract came under heavy attack by Congressman Ron Wyden of Oregon and the NIH, which supplies $70 million of the $120 million in Scripps's annual budget, a scaled-down version of the agreement was fashioned in which Sandoz is to provide $100 million over five years in exchange for a maximum of 47 percent of Scripps's inventions and a reduced period of 90 days in which Sandoz may exercise its options. In addition, Scripps agreed to create an Office of Small Business Collaborations to ensure that small enterprises will also have access to the federally funded discoveries.

With severely declining budgets, entire university systems are being pressured by legislators, trustees, and alumni to link academic research to business enterprise. The unfortunate consequences of having medical schools engaged in a highly competitive business have been plain to other sectors of the universities. Salaries for professors in lucrative clinical specialties reach the million-dollar mark, and their money-making capacity becomes a determining factor in choosing the leadership of a department. Now, with the laudable intention of accelerating technology transfer, scientists and engineers throughout such universities are being urged to follow the example of the medical school.

Early in 1993, at the University of California, the largest and most outstanding state university system in the country, two corporations were to be created to foster technology transfer. One was a nonprofit corporation, the UC Technology Development Foundation, that would offer inventors legal, technical, and financial services and would retain the patents as an agent of the university. The other, the UC Technology Development Company, was meant to be a profit-making enterprise that would nurture promising inventions and discoveries with seed money, enabling them to grow into start-up companies to be sold to private investors, with the options to retain a share of the equity. Expecting that twenty start-up companies would direct 100 inventions to the marketplace each year, it was expected that funds in excess of $222 million would be generated by the year 2001.

To meet such heady financial targets, the university administrators would have had to reorder traditional attitudes that have accorded to intellectual achievement and to devotion to teaching a status well above that of marketing. In the words of one top UC official, "A substantial number of faculty members are genuinely scared. Then there is a small group of faculty members who are genuinely excited about the change, because they see that this is a vindication of what they had believed for many, many years, namely that university research must also be relevant research."

Some months after UC officials had announced their ambitious plan to commercialize UC science, they were forced to withdraw under a barrage of faculty criticisms, including questionable violations of state law and university ethics policies and unusual associations with business executives in line to profit from UC projects. Added to these anxieties, some of the boosters of the plan were already crafting generous corporate-style pay and benefits packages for themselves, ostensibly for their superior management of the university.

A year after the UC initiative, Columbia University announced its own, The Columbia Innovation Enterprise, a "bridging organization" to implement university research for industrial use. The Enterprise, according to press accounts, is to make "a conscious, planned effort" to help Columbia researchers modify their inventions to render them "more palatable for outside consumption." To maintain its balance on the slippery slope of faculty–industry collaborations, the university will rely on existing conflict-of-interest rules and a prohibition of owning shares in companies for which an academic is doing sponsored research.

Among the many concerns for academic institutions in promoting patents and industrial partnerships, the most serious are the illusions fostered by the few instances in which an individual or department appears to have profited greatly. On closer examination of each case, the financial gain represents only a small fraction of the total income needed by the institution to sustain its research and training missions. Another illusion is that the gap left by the grossly inadequate budgets of the NIH and the NSF can be filled by industrial initiatives and corporate contributions. Not only would sums from such sources yield only a small percentage of the overall need, but the conditions for this support would tend to direct the research too narrowly toward the development of the company's products. Still, we can expect that the increasing need to sustain training and research programs on inadequate budgets will increase the resonance of the notion of relevance in research. Inevitably, the reverberations of focused research will tend to suppress imaginative thinking about the arcane questions that (it is hoped) will generate the major revelations of the future.

Biotech Ventures: Pros and Cons

The astonishing technological advances based on genetic engineering—the most profound in the history of biology and medicine—came entirely from academic institutions ill-equipped to develop discoveries into marketable products. This breach was filled by biotech ventures with members of university faculties involved in their founding and direction. These entrepreneurial connections entail advantages and threats to the progress of science and human welfare that have already been considered and need only be mentioned again.

On the pro side, biotech companies provide the most effective means of translating advances in science and technology into products for medicine and agriculture. This crucial function is performed less well by the large pharmaceutical companies and virtually not at all by academic institutions. At the forefront of cloning genes and producing key hormones, cytokines, and receptors for medicine, biotech ventures provide these powerful reagents and emerging insights for the benefit of basic research in biology and chemistry.

Currently, biotech ventures inject more than one hundred thousand high-level jobs and many billions of dollars into the economy, and these numbers are likely to expand manyfold by the year 2000. Among these jobs are a variety of career opportunities for biologists and chemists. The increase in the volume of scientific activity generated by biotech ventures promotes and expands satellite industries that innovate equipment and reagents and effect major economies in their cost. The affluence gained by some scientists associated with biotech ventures generally engenders respect in the business world and creates a positive image for biological science that links financial success with practical values.

Most impressive is the focused, communal effort that can be found in a biotech venture, an ambience that is difficult to generate or to sustain in academia. This deficiency in the organization of academic research is perhaps the only unfortunate consequence of the NIH- and NSF-supported programs that have revolutionized the quality and magnitude of medical and biological research in the post–World War II era.

An academic investigator, young or old, judged worthy by a remote committee of scientific peers assembled by a federal agency, is granted direct and complete control of a research program, free from any constraint or intervention by a departmental chairman or by a university dean or administrator. A consequence of this independence is a lack of incentive for interaction with other research groups in the same department or on the same university campus. Generally, the research unit of five to ten or more people is a small duchy, self-contained and self-sufficient. Each unit needs to identify its own achievements in order to remain competitive in attracting students and in securing the grants that keep it viable. Grants are also given by the NIH to groups of investigators to encourage cooperative and interdisciplinary research, but such groups often prove to be little more than paper organizations; despite diligent efforts, team spirit and joint actions are difficult to maintain.

The biotech venture can provide an alternative research setting with certain advantages, not only over academic institutions, but also over industrial laboratories. Encumbered by multilayered bureaucracy and by broad and diffuse missions, industrial research is pursued in an atmosphere strongly affected by problems of manufacturing, regulatory controls, marketing strategies, and the need to turn a profit. Yet the large pharmaceutical companies are not burdened by such constraints as those placed on academia by the NIH and the NSF funding patterns; they can innovate by decentralizing their research enterprise into smaller, dispersed units. Entities the size of DNAX, insulated from the commercial distraction of the parent company, but integrated into its goals and operations, can provide dynamism and better focus on prospects for drug discovery.

On the con side, biotech companies are not in business primarily to do research and to acquire knowledge for its own sake; rather, they are in research to do business and to turn a profit. They possess neither the mandate nor the tradition to advance scholarship. Ventures must instead prove their profitability in the ebb and flow of financial markets and despite abrupt changes in corporate management that may place greater emphasis on short-term goals. With their successes, some venture capitalists have assumed heroic dimensions at the expense of those who achieved the scientific advances, although it must be admitted that entrepreneurs who risked and failed often took the blame and ended up as goats.

Profit-driven ventures are more likely to promote secrecy and to subvert academic units with patent agreements and joint ventures that demand exclusive access to discoveries. Scientists, students, and resources dedicated to the ventures are diverted from basic, untargetted research to projects that are too highly focused and selected for the prospect of early profitability.

A more subtle con is that the publicized vigor and successes of biotech companies may foster illusions that basic research can be left to industry. In fact, more than 90 percent of such research has been, in the past, and must be, in the future, done in university and other academic settings, requiring massive support to the tune of billions of dollars from the taxpayer through the federal government.

Organized, sustained, and diligent efforts are needed to glamorize and fund basic research and to emphasize how essential such long-term investment is to solving the major problems in medicine. The source of the greatest practical advances in medicine—x rays, laser surgery, and MRI; polio vaccine, recombinant DNA, and monoclonal antibodies—have all come from the pursuit of curiosity in basic science, initially without any thought of a defined commercial objective. The decades of painstaking, fundamental work in academic institutions needed to make these practical applications possible was not supported by an industrial organization, nor is it likely that it would be in the future.

When it comes to the *application* of basic discoveries, we must rely on the pharmaceutical industry for the refinement and mass production of vaccines, hormones, and drugs, for the perfection of instruments and diagnostic procedures, and for the intricate processes of quality production, regulatory approval, and responsible marketing. We can and must look to industry for the large-scale operations and the expert management of a myriad of difficult problems for effective drug development and distribution.

The balance between the good and bad features of biotech companies varies from one to another. I have been fortunate that my associations with them over the past 25 years have been instructive and stimulating, both at the scientific and at the personal level. Having been a member of the scientific advisory boards of ALZA, DNAX, Regeneron, XOMA, and GalaGen, I find that they differ in style and organization, but all serve science and society in positive ways.

Lessons from Biotech Ventures

The pros and cons of biotech ventures, after nearly two decades of experience and hundreds of variations, have important lessons to teach—not only about success in such ventures, but also about the pros and cons of medical research in pharmaceutical companies and in academic institutions. The criteria of success will differ

widely, depending on the weight given *science, profit,* and *integrity.*
Weakness of any member of this three-legged stool that supports
a biotech venture makes for, in my view, a wobbly enterprise. In
general, the same can be said for large companies and academic
enterprises.

First, the science. Even in this computer age, science is still done
by people, not by formula and not by machines. What matters most,
in the long haul, is the quality of the scientists. Bold and vigorous
explorations in science require artistic temperaments and artistic
judgments, not unlike those that influence writers, painters, and
composers. The tools in science are more technical and precise, but
the goal of probing the secrets of nature is much the same. The
scientist must strive hard to produce something important and,
within the immediate community, to be eager to share ideas, facts,
and reagents and to derive personal pleasure in the pursuit of common goals. As in athletics, a team's achievement is more than the
sum of individual efforts, and an outstanding player—Arai at
DNAX, Goeddel at Genentech, Yancopoulos at Regeneron—can
be a driving and cohesive force.

To get and to keep the best and truest scientists requires an environment in which the resources do not limit productivity and
creativity. In most enterprises, industrial and academic, the facilities, supplies, and technical help are adequate. What is most generally lacking is a climate of openness. Secrecy is the enemy. Free
communication inside the unit and with the scientific community
at large insures that personal achievements will be recognized,
which is the only true tenure for a life in science. Secrecy thwarts
not only personal recognition but also the free trade of scientific
knowledge that builds bridges between industry and academia to
the profit of all.

The scientific vigor of a biotech venture may diminish with
growth in size and with a scientific staff that has been in place for
some years. Interests and directions diffuse, groups become insu-
lar, and the initial excitement of finding and achieving goals may
wane. A strong postdoctoral fellowship program and a dedicated
scientific advisory board can help sustain momentum and preserve
high standards.

In the absence of turnover and the constant flow of students
found in academic departments, energetic postdoctoral fellows do
much to enliven the research atmosphere of a venture or pharmaceutical company. Recruitment of excellent fellows to a venture

has been made easier by virtue of the decreasing number of fellow-ships made available in academia and by the increasing attractive-ness of careers in industry. Because of their modest salaries and benefits and their brief tenure, fellows impose far less strain on closely monitored budgets and head counts. Assignment of a fellow recognizes achievement by a staff scientist and enhances stature be-yond financial awards. Fellows bring fresh approaches and tech-niques from their varied academic backgrounds, they challenge entrenched ideas, and they constitute a pool of potential staff can-didates. On leaving, fellows can become loyal, active alumni, who may direct a stream of potential fellows to the enterprise.

A scientific advisory board more than pays for itself. The effort and cost required to assemble its half dozen or so members twice a year is rewarded by the traffic it brings across this bridge to aca-demia. The scientific staff, while enjoying the freedom of not hav-ing to prepare competitive grant applications, is also deprived of the salutary stimuli and pressures that accompany periodic appli-cations and reviews. The advisors provide an alternative to peer-review panels. In this capacity, they judge the quality and directions of the research and the skill and effectiveness of the scientists who appear before them. In their presentations to the board, the sci-entists and their research gain visibility and recognition and en-gender closer ties with the academic community. The board may function also as a resource through which staff and postdoctoral fellows may be recruited, as a wellspring of counsel for the man-agement, and as a source of encouragement for investors.

The second of the criteria that are important to the success of a venture is that it be profitable, at least profitable enough to sus-tain its vitality. For this, skillful managers are needed to conduct the business operations. Understanding and support of the science must be coupled with a talent for managing people and a talent for the pursuit of financial arrangements. Until such time as products, royalties, and contracts return a regular profit, various modes of financing must be acquired, and joint ventures must be created, to cover the cost ("burn rate") of the growing enterprise and to pro-vide a comfortable margin of safety.

The CEO is the key to the whole operation of a biotech venture, an alchemist who can turn DNA into gold. This slot is the most dif-ficult to fill. Venture capitalists, who typically have a strong voice in the early organization of a biotech venture, generally demand that the CEO be someone with business training and experience. In doing so, they may discount or ignore any latent entrepreneurial skills of the scientists who helped to launch the venture. The remarkable business talents and achievements of such scientists as George Rathmann at Amgen, Bill Rutter at Chiron, and Len Schleifer at Regeneron were described in Chapter 7. There are many other examples, including Laurence J. Korn, who founded Protein Design Laboratories (PDL) in Palo Alto.

Korn, a native New Yorker, took an undergraduate degree in biochemistry at the University of California, Berkeley, and a Ph.D. in molecular biology with Charles Yanofsky at Stanford. After prestigious postdoctoral fellowships with Donald Brown at the Carnegie Institution in Baltimore and with John Gurdon at Cambridge University, he accepted an assistant professorship in the Genetics Department at Stanford. After six years (1981–1987) in that department, he had grown tired of the wrangling and divisiveness. He invited Cary Queen (a Cornell mathematician turned molecular biologist, then working at the NIH, who had been a friend since high school) to join him in founding PDL. Their mission, which was to design antibodies for the treatment of autoimmune diseases and cancer, attracted lucrative joint ventures with Roche, Sandoz, and Yamanouchi; an initial public stock offering netted $50 million in 1992. A year later, Korn concluded the largest corporate alliance in the history of biotech ventures that did not involve a change in ownership. For $206 million, Boehringer-Mannheim acquired 15 percent of PDL equity and marketing rights in Europe for several products already in final (phase III) clinical trials. During its seven years of extensive growth, PDL incurred a net loss of only $10 million, which is remarkable for a biotech start-up.

The third of the criteria that are important to the success of a biotech venture is integrity: sound business and scientific principles and decency in their execution. Keeping a biotech venture financially alive to achieve its product goals can strain standards of integrity. Too often, investors and Wall Street advisors are attracted by glitz and jackpot potential, and they turn away from strong science, with its slow pace and its conservative predictions. The objectivity of science and its record of progress, coupled with sound management, offer the best security for profitable investment.

In the biotech world, with vast opportunities for innovations in science and untold applications for improving the human condition, good guys needn't finish last. Under the pressures of intense competition and the imminence of failure, the temptation to cut corners and thereby to gain an advantage over an apparent rival may be ever present. Yet, whether in the discharging of professional responsibilities or in the conduct of personal relationships, there need be no concessions to conflict of interest, to gimmickry in marketing, to litigiousness over patents, to profiteering, and, over all, to a lack of commitment in supporting basic science and in contributing to community welfare.

A much-publicized issue of integrity is the conflicted interest of an academic scientist engaged in entrepreneurial or consultative roles in biotech ventures. Academic research designed to aid a venture is inappropriate, as is privileged access to academic research given to a venture. Entirely appropriate is the devotion of a fraction of an academic's time—one day per week is a long-standing Stanford University rule—to participate in founding a venture and directing its efforts toward discoveries and applications with resources and a focus that could not be created in academia.

Of arguable propriety is the widely practiced university–industry venture. In 1990, 1058 such arrangements on 203 campuses had research-and-development budgets totaling $2.66 billion, a sum to be compared with NSF research outlays of only $1.69 billion in the same period. A survey by Richard Florida of Carnegie-Mellon University disclosed that, in roughly half of the programs, research information was delayed or withheld in order to favor the sponsors. It should be of great concern whenever an academic institution accepts industrial support for basic research that is specified by the industrial donor, especially when the rights to any discoveries made in that research are reserved for the donor.

Marketing is essential to making the world aware of a product and its value. Although one can criticize the allocation to marketing of resources that are huge when compared with those allocated to research and development, such choices properly remain within the province of business management. What I find objectionable and deceptive is the gimmickry of attaching a trademark to an age-old pharmaceutical and marketing it as something novel in substance and efficacy. An example among many is Tylenol, one of 58 commercial disguises for acetaminophen. This simple chemical was synthesized back in 1878, and its capacities for relief of pain and fever were well documented in numerous reports over the decades. While lacking the anti-inflammatory actions of aspirin, as well as some of aspirin's other beneficial effects, acetaminophen has the advantage of not causing some of aspirin's side effects, such as gastric bleeding and discomfort. Tylenol brought the McNeil division of Johnson & Johnson $1.05 billion in sales in 1993. The price charged to the customer can be ten times that for the selfsame substance, on the same shelf, labeled No-Aspirin. Beyond the excessive cost is the illusion of special potency in *Extra-Strength* Tylenol, which is merely a larger pill containing 500 milligrams of acetaminophen instead of the customary 375. Once, when I asked my druggist for a reliable generic acetaminophen rather than the far more expensive Tylenol, he told me that he stocked only the superior Tylenol because of the extensive and costly research that he believed went into its formulation and production.

Not uncommonly, the race for patent and regulatory approval of a drug appears too close to call. To avoid protracted and costly litigation, cross-licensing agreements give each party latitude and incentives to discover improvements and novel uses of the drug. Such, for example, was the agreement between Schering-Plough and Roche to share alpha interferon. By contrast, a winner-take-all approach, with the intent to destroy a financially weaker rival, can prove disastrous. Centocor Corporation, in pressing a 2½-year legal battle with XOMA Corporation over the patent for antibodies to treat septic shock, nearly destroyed both biotech companies, retarded development of the drug, and depressed investor confidence in the whole biotech industry.

"Charge what the traffic will bear." Whether this dictum originated in the pharmaceutical industry is debatable, but price gouging, as in all industries, has surely occurred. In some instances, in the 1980s, a higher price was sometimes the chief distinction in marketing between one drug and another lackluster "me too" drug. Yet a unique and life-saving drug is worth a high price. The brevity of the monopoly enjoyed by a drug, the high cost of developing it, and the frequency of failures can justify the high price. Instances in which companies have succumbed to the temptation to overcharge, on top of the profitability of the industry as a whole, have led to charges of widespread abuse and demands for price controls. Aside from the poor record of price controls in any industry, punitive measures are likely to discourage the high risks of investing in new drug discovery and development, and they may lead to even higher costs for stagnant, ineffective health care. To the late Mary Lasker, a great evangelist for medical research, we owe this aphorism: "If you think research is expensive, try disease."

The partnership of biotech ventures and pharmaceutical companies with basic biomedical science in academia needs constant attention and nurturing from all sides, including the public and its governments. The health and wealth of society depend on it. We need to be aware of what is owed by one sector to another in order that each may enhance the fulfillment of its capacities and responsibilities.

The industrial sector must be reminded that the concepts, techniques, and practitioners of biotech ventures and biotechnology-driven pharmaceutical companies all came from basic academic research. As technologies proliferate and gain in sophistication, their scientific foundations are submerged and forgotten. The competitive rush to the marketplace leaves little time and no resources to ensure that the foundations for novel technologies will be built for the next generation.

The academic sector must keep in mind that the industrial enterprise converted their scientific advances into the reality of products for improving human welfare. Industry gave employment to scientists, and it developed the technologies and equipment that accelerated basic science and strengthened public sympathy for it by contributing the visible benefits that accrue from research.

Most important of all, the public must be informed that the viability of the academic and industrial sectors of the biomedical enterprise depends on their understanding and support. To this end, both partners must maintain a strong and constant chorus of information to the public and to government about their impressive achievements, which made the United States the world leader in biomedical research and innovation. Serious diseases now demand a strong and sustained commitment for their solution. Long-term support is essential for academic science to increase our understanding of the basic chemistry and biology of living matter. For industry, incentives must be available to entrepreneurs to invest in the risky business of applying new knowledge to discover the drugs and devices we need to prevent and treat disease.

The Future of Biotech Ventures

Academic discoveries in molecular biology, genetic chemistry, and related disciplines provided the foundation for industrial development of novel and important drugs and devices for medicine and agriculture. Although the basic knowledge was obtained largely in the United States, the information and techniques embodied in biotechnology spread rapidly around the world. What is uniquely American is the massive number of biotech ventures. In 1993, more than 1200 such ventures employed nearly 100,000 people, and perhaps again as many were employed in high-skill jobs in supporting industries. Forecasts predict that this $6 billion industry will expand severalfold by the year 2000. Will this vaunted growth take place, or, as with other failed economic predictions, might there be a reversal instead?

The impact of the genetic-engineering revolution is most visible in the exciting new drugs, vaccines, and diagnostic procedures already being marketed or known to be in the pipeline. Yet, the prospects are far brighter in the opportunities that have been afforded by basic research to advance the understanding of all life processes in molecular terms, expressed in the unambiguous language of chemistry. These advances will range from the mapping and sequencing of genomes—including those of microbes, mice, and humans—to fundamental insights into the nature of growth, development, and behavior and their aberrations in disease and aging.

As our knowledge expands, so do the questions. To pursue these new and important questions, we will need novel principles and techniques created by trained and talented scientists. Vigorous support of basic research and training from federal sources is imperative to sustain the current momentum and to sow the seeds for the certain but unforeseeable advances of future revolutions. Despite the importance of funding basic research, the most immediate concern in the growth of our biotechnology enterprise and the viability of existing ventures is private investment.

Biotech ventures are ignited by venture capitalists and fueled by additional private investments and public stock offerings. In every instance, what the investor looks for is a quick and substantial profit. When Regeneron prepared its initial public offering, the expressed interest among prospective buyers ran so high that the company and underwriters could select from among them those institutions and individuals who had professed intentions for long-term investment. Yet, within the first day of trading, half of the shares had turned over.

The risk of investment in biotech ventures is high. Fueled by greed and optimism, the ventures expand in unreasonable numbers. They pay no dividends, and, with rare exceptions, they sustain losses year in and year out. The gamble is on growth and profit in the future. The venture capitalist expects to multiply his investment, perhaps tenfold, when the company is taken public in two to three years; he can afford five failures for each success, but not ten. Success of a biotech venture depends on a profitable drug. Whereas the large drug company can pursue twenty or more drug-development projects at any given time, the biotech venture must stake its future on only one or two. When a project fails, as most do, the large company buries the corpse without any notice, while the biotech venture must conduct the funeral in the glare of the Wall Street press.

The public investor hopes for a bonanza, like Genentech or Amgen, but the typical biotech venture languishes until acquired by another start-up or by a major pharmaceutical company. Maturation of a venture into a fully integrated company with proven manufacturing, regulatory, and marketing capabilities is exceedingly unlikely, and it has been achieved to date only by Genentech, Amgen, and Chiron. Even for those successful ventures, future independence is uncertain, in view of intense competitive pressures that have, in recent years, driven mergers of company giants, such as Squibb with Bristol-Myers, SmithKline Beckman with Beecham, Searle with Monsanto, Marion with Dow, Parke-Davis with Warner Lambert, Rorer Group with Rhone-Poulenc, Medco Containment Services with Merck, and Syntex with Roche.

Added to these daunting problems are charges of gouging and plans for federal control of drug prices, a thrust that has already discouraged investments in pharmaceuticals; the chilling effect has been even greater on the more fragile biotech ventures. Were such disincentives to profitability imposed, these nascent ventures might well be frozen to death.

The staggering cost of health care and the expense of the newer drugs have given rise to several misconceptions. Few people realize that drugs consume only 6 to 7 cents of the health-cost dollar, only a small fraction of what is spent on administration, hospitals, and doctors' fees. Even if drug prices were fixed at their present levels, the savings in health costs would be negligible.

To bring a new drug to market now takes, on average, more than ten years, at a cost in excess of $300 million and with an expected market lifetime of relatively few years, owing to competition from new technology and the limited patent life. In addition to the cost of developing a marketable drug are the far more frequent failures. Of four thousand compounds screened in preclinical testing, only five make it to clinical trials, and only one of these is finally approved. Even the very costly drugs are still economically sound in sharply reducing hospital stays and prolonged treatment, thus enabling the patient to return earlier to gainful employment. Drugs eliminated costly ulcer surgery, reduced mortality from hypertension and infections, and made possible ambulatory care for mental illnesses, kidney disease, and cancer. Beyond cost effectiveness, and not to be ignored, is the improvement of the quality and duration of life that these drugs brought to the treatment of previously intractable diseases.

The future of medicine and the pharmaceutical industry depend on innovation—new drugs that will cure the currently debilitating and degenerative diseases, such as AIDS, Alzheimer's disease, multiple sclerosis, most cancers, rheumatoid arthritis, and assorted genetic disorders. A Boston Consulting Group study, recently released, contends that the annual medical costs of just seven of the major uncured diseases account for half of the current health-care bill. Without innovation, we will be mired in the current state of high-cost, ineffective treatment.

Cost restraints and competition will increasingly discourage the introduction of drugs of the "me too" variety and will place a premium on the novel. The discovery of these truly unique and effective agents will more likely emerge from small, highly focused operations, such as biotech ventures, than from huge, multitiered research divisions of the large pharmaceutical companies. The serious danger of faltering investment in biotech ventures is that many will either fail or fall prey to "cherry-picking" by the pharmaceutical giants. In 1992, drug companies entered into about fifty strategic alliances with biotech firms; there were twice as many in 1993, and the prospects for 1994 are for an even greater number. Clearly, the preservation of the biotechnology industry, as practiced by the start-ups as well as by the big companies, is in the economic and health interests of the nation, and that industry must be protected against threats to its financial viability.

A wealth of knowledge and technology is available to biotech ventures and to the pharmaceutical industry for immediate application to the development of drugs for the prevention, diagnosis, and treatment of some diseases. Yet this fund of information is woefully inadequate for the far greater number of diseases that need to be confronted. This new knowledge depends on research that explores the basic biology and chemistry of a wide range of cells and organisms. More than 90 percent of the support for such basic research must come from the NIH, which, in the past four decades, has been largely responsible for the emergence of biotechnology and other major advances in medical science. Notable economists and politicians who attack federal spending for basic research and go so far as to propose privatization of the NIH are simply unaware that a long, arduous, and expensive investment in basic science, which no industry could possibly have afforded, was responsible for the medical miracles that they and their families enjoy. They do not appreciate that basic research has been, and will remain, the very lifeline of medicine.

In the last analysis, the pharmaceutical industry and academic institutions will weather change and survive. Of the biotech ventures, only a very few will mature into independent companies, and a few others will be assimilated into existing pharmaceutical giants; the vast majority will likely disappear. This would be most unfortunate, because biotech ventures bring an intense thrust to an important area of science and a commitment to application that academia cannot supply and that the huge pharmaceutical company cannot do nearly as well. Biotech ventures, in unique and crucial ways, make precious contributions to science and to the health of nations.

Epilogue

The focus of this narrative has been on the DNAX–Schering-Plough venture and, more particularly, on the people who had the vision and courage to start biotech ventures and the motivation and skill to make them succeed. Because the characters in real-life dramas must change, one should look to institutional practices and traditions for keys to the future. Exuberance fueled by youthful success should be tempered with concern for maintaining that vigor during maturity. A few biotech ventures may develop into stable, profitable companies, but one must wonder whether they or DNAX can sustain a focus on basic research when faced with constant scrutiny of its cost-effectiveness. Even venerable institutions with enviable records of achievement—Bell Labs, General Electric, Xerox, and IBM—have, in recent years (and for various reasons), sharply curtailed their support of basic science.

Schering-Plough—a pharmaceutical company in a convulsive health-care industry—will be forced to examine every detail of its costs and may be hardpressed to justify long-term investments in basic research at DNAX. This accounting will be especially likely should a downturn in business coincide with a slackening in research productivity. At such a juncture, I would hope that wise managers will reflect on what DNAX has achieved over the long haul in supplying novel drug candidates, upgrading the quality of research throughout the company, and improving the company's image.

There are lessons from the DNAX–Schering-Plough experience both for industry and for academia. For a large pharmaceutical company committed to the discovery of novel drugs, the DNAX model—a small institute, remote from the imperatives of product development and marketing and given the freedom to innovate—is a sounder investment than a huge company laboratory of the sort now in place at most established companies. However,

259

geographical and cultural separations require that personal inter-actions between the institute and the parent company be cultivated and maintained at every level.

For the university concerned with pursuit of knowledge and its transfer to promote human and economic welfare, the DNAX achievements demonstrate that basic science can be pursued quite effectively in another setting, and they should direct attention to the shortcomings inherent in the operations of large academic in-stitutions that impede the creativity of their faculties. Finally, the DNAX–Schering-Plough relationship illustrates that, in the long term, the scientist and the industrialist, with mutual confidence and trust, can produce good science and sound business of signif-icant social merit.

References

Angier, Natalie. 1989. *Natural Obsessions: The Search for the Oncogene*, pp. 283–284. Boston: Houghton Mifflin.

Campbell, Russ, editor. 1991. *The Dynamics of Silicon Valley*, pp. 34 and 91. San Jose, Calif.: Western Business Forum.

Djerassi, C. 1992. *The Pill, Pygmy Chimps, and Degas' Horse*. New York: Basic Books.

Hall, Stephen S. 1988. *Invisible Frontiers: The Race to Synthesize a Human Gene*. New York: Atlantic Monthly Press.

Harvey, S. 1982. Alex Zaffaroni: Drug-delivery pioneer. *Pharmaceutical Executive* 2(11):26–30.

Herzog, H., and E. P. Oliveto. 1992. A history of significant steroid discoveries and developments originating at the Schering Corporation (USA) since 1948. *Steroids* 57:617–623.

Kornberg, A. 1989. *For the Love of Enzymes: The Odyssey of a Biochemist*. Cambridge, Mass.: Harvard University Press.

———. 1991. *It All Began from a Failure*. Japanese translation of *For the Love of Enzymes* by Kenichi Arai, Hisao Masai, and Yoko Nakayama. Tokyo: Yodo-sya Publications.

Syntex Laboratories. 1966. *A Corporation and a Molecule: The Story of Research at Syntex*. Palo Alto, Calif.: Syntex Laboratories.

Weissmann, C. 1981. The cloning of interferon and other mistakes. *Interferon* 3:101–134.

Zaffaroni, A. 1953. Micromethods for the analysis of adrenocortical steroids. *Recent Progress in Hormone Research* 8:51–83.

———. 1978. Industrial development of controlled delivery drug systems. In *Proceedings of General Electric Company International Symposium on Science, Invention, and Social Change*, pp. 73–80.

———. 1988. Future perspectives on therapeutic systems. In *Visions and Values for Pharmaceutical Innovation*, pp. 151–174. Palo Alto, Calif.: ALZA Corporation.

———. 1992. From paper chromatography to drug discovery. *Steroids* 57 (12):642–648.

Glossary

acetic acid: The simple compound CH_3COOH, known as vinegar in the dilute form, is a common intermediate in cellular metabolism.

acetyl coenzyme A: The activated form of acetic acid in cellular metabolism.

acquired immunodeficiency syndrome: *See* AIDS.

actinobacteria: Bacteria of the phylum Actinobacteria, commonly known as actinomycetes. Included in this phylum are bacteria of the genera *Micromonospora*, *Streptomyces*, and *Mycobacterium*, the last of which includes the species responsible for leprosy and tuberculosis.

adenosine triphosphate: ATP, the major currency of energy transactions in the metabolism of all cells.

adjuvant: A substance that, when injected together with an antigen, increases the immune response of an animal.

AIDS: Acquired immunodeficiency syndrome, the fatal disease caused by HIV (*which see*) that breaks down the immune system's defenses against a variety of microorganisms.

aldosterone: The major adrenal steroid hormone regulating mineral metabolism in humans.

alpha interferon: IFN-α, one of the large family of proteins (cytokines) released by leukocytes (white blood cells). Among their multiple regulatory functions are interference with viral multiplication and with the proliferation of cancer cells.

ALS: Amyotrophic lateral sclerosis, a fatal disease of the nervous system. Also known as Lou Gehrig's disease, after the afflicted baseball star.

Ames test: A bacterial bioassay for detecting mutagenic compounds; developed by Bruce Ames in 1974.

amikacin: A semisynthetic aminoglycoside (*which see*) antibiotic derived from kanamycin A and used in the treatment of bacterial infections.

amino acids: The building blocks of proteins. Twenty distinctive amino acids (e.g., serine, tryptophan, glutamic acid), in some particular sequence of one hundred or more, constitute all cellular proteins.

aminoglycoside: One of a group of antibiotics (including streptomycin, kanamycin, and neomycin) whose action is to inhibit protein synthesis in bacteria.

amyotrophic lateral sclerosis: *See* ALS.

antibody: A protein formed by an animal in response to invasion by a foreign substance (e.g., bacterial, viral, plant) that binds with that substance and neutralizes its action.

> **anti-idotypic:** An antibody formed in response to the presence of another antibody.

> **catalytic:** An antibody designed by chemists to display enzyme-like (catalytic) properties, as well as to retain the high specificity of recognizing and binding the foreign agent (antigen) that induced the formation of the antibody.

> **monoclonal:** An antibody derived from a single clone of cells. All of the molecules of such an antibody are therefore chemically and structurally identical.

> **polyclonal:** A population of antibodies derived from many clones of cells, each producing antibodies that are directed against different facets of the foreign antigen. Although they are very similar, the antibodies are chemically and structurally different.

antidigoxin: An antibody directed against digoxin, a toxic steroid compound derived from the foxglove plant (*Digitalis purpurea*) that is used in the treatment of heart disease.

antigen: A substance that can stimulate an animal to produce antibodies that can then bind the antigen. Generally, an antigen is a foreign substance, but, when it is produced by the same individual that produced the antibody, it can be the cause of disease.

arthritis, rheumatoid: An inflammatory joint disease that is thought to be due to autoimmune responses.

ATP: *See* adenosine triphosphate.

autoimmune disease: Any of a group of diseases (e.g., juvenile diabetes, lupus, rheumatoid arthritis) due to the formation of antibodies against antigens produced by the same individual.

bacteriophage: A virus that infects bacteria.

B cells: Lymphocytes formed in the bone marrow of mammals that function in the immune system to produce antibodies both in the blood stream and secreted onto mucous surfaces.

BDNF: Brain-derived neurotrophic factor, a protein released from brain cells that promotes the growth and maintenance of certain nerve cells.

beta interferon: IFN-β, a cytokine released from fibroblasts (*which see*).

Betaseron: A trade name for beta interferon (*which see*).

bioassay: A measurement of either the activity or the amount of a substance that is based on the use of living cells or organisms.

biological: The noun denotes a product derived from cells or organisms.

bleomycin: A group of related glycopeptides (*which see*) isolated from *Streptomyces* that cause strand breakage in DNA and that are used in the treatment of some cancers.

blood–brain barrier: A cellular barrier in the brain that prevents the passage of many substances from the circulating blood that readily penetrate most other tissues. The barrier serves to minimize fluctuations in the level of substances that might interfere with the constant environment needed for brain functions.

B lymphocytes: *See* B cells.

brain-derived neurotrophic factor: *See* BDNF.

catalysis: An increase in the rate of chemical reaction brought about by the action of a catalyst, generally an enzyme.

cDNA: Complementary DNA, a DNA molecule produced by an enzyme (reverse transcriptase) using RNA as a template. cDNA is used in genetic engineering to identify and clone the genes that encoded the RNA.

charge-transfer complexes: Complexes formed in cellular oxidation–reduction reactions when the transfer of electrons bonds the positively charged donor to the negatively charged acceptor.

chemokine: One of a family of small proteins (cyto*kines*), functioning as *chem*ical attractants, that promote cellular interactions.

Chlor-Trimeton: A trade name for the antihistamine drug chloropheniramine.

choline: A simple compound that serves as a constituent of phospholipids (in cell membranes) and (combined with acetic acid, as acetylcholine) in the transmission of nerve impulses. It is required as a dietary supplement under some conditions.

chromatography: The separation of complex mixtures of molecules on paper or resins.

chronic bowel disease: Serious inflammatory diseases of the bowel (e.g., chronic ulcerative colitis, Crohn's disease).

ciliary neurotrophic factor: CNTF, a protein that promotes the growth and maintenance of certain nerve cells.

clonal selection process: The means whereby a particular cell clone is selected among a very large number of clones. In the case of lymphoid cell clones, the binding of an antigen to a specific, genetically determined receptor on the cell surface stimulates that cell to multiply to the virtual exclusion of other cells that lack that receptor.

CNTF: *See* ciliary neurotrophic factor.

coenzyme: A small nonprotein molecule that, when tightly bound to a large protein molecule, converts it to an active enzyme molecule.

colitis, ulcerative: *See* chronic bowel disease.

colony-stimulating factor: CSF, a family of proteins (cytokines) that stimulate the growth and development of hemopoietic cells. For example, G-CSF (*which see*) is the factor that stimulates the growth and development of granulocytes, and it is used in the treatment of disorders in which the circulating blood level of these essential cells is depressed.

complementary DNA: *See* cDNA.

complement fixation: Activation of complement, a group of nine or more proteins in the blood that complement the actions of antibodies in killing cells.

corticosteroid: A steroid produced by the adrenal cortex that may serve as a hormone (e.g., regulating sexual function or mineral metabolism) or as a source of cholesterol, a nonhormone.

cortisol: A major steroid hormone derived from progesterone possessing strong anti-inflammatory activity.

Crohn's disease: *See* chronic bowel disease.

cytokines: A group of hormone-like proteins released by some cells to stimulate others. They include interferons, interleukins, colony stimulating factors, growth factors, and tumor necrosis factors. Cytokines produced by lymphocytes are called lymphokines.

deoxyribonucleic acid: *See* DNA.

dideoxynucleotide: An analog of a deoxynucleotide (a building block of DNA) that lacks a hydroxyl group whose presence is essential for synthesizing a DNA chain.

Dig-Annul: A proposed trade name for an antibody to be used as a drug to neutralize toxic levels of digitalis and its derivatives.

diosgenin: The component of the steroid used by R. E. Marker in the synthesis of progesterone.

DNA: Deoxyribonucleic acid, the macromolecule of heredity in virtually all organisms. The particular sequence of the four deoxynucleotide building blocks of DNA (A, T, G, and C) spells the identity of a gene.

    complementary: *See* cDNA.

    recombinant: DNA, created generally by genetic engineering, in which genetic sequences from diverse origins are joined together (recombined) to form a novel molecule.

DNA polymerase: The enzyme that catalyzes the assembly of deoxynucleotides into DNA guided by the sequence of the building blocks of DNA (abbreviated A, T, G, and C) in a preexisting DNA chain serving as a template.

EBV: Epstein-Barr virus, the herpesvirus that causes infectious mononucleosis and that is implicated in Burkitt's lymphoma.

ecdysone: A steroid hormone in insects that regulates metamorphosis (e.g., molting of caterpillars, pupa formation).

electron spin resonance: ESR, the common name for electron paramagnetic resonance. Spectroscopic measurements of ESR identify free radicals in organic compounds.

enzymes: Protein molecules that catalyze virtually all the chemical reactions in living cells. Each enzyme is highly specific for the conversion of a particular molecule (the substrate) to a particular product.

EPO: Erythropoietin, the cytokine produced in the kidney that is essential for a stage in the development of red blood cells (erythrocytes) in the bone marrow.

Epstein-Barr virus: *See* EBV.

erythropoietin: *See* EPO.

eukaryote (or *eucaryote*): An organism (unicellular or multicellular) of a higher order than prokaryotes (such as bacteria) and distinguished by having cells with a nucleus surrounded by a membrane and internal structures, such as mitochondria.

expression cloning: A technique for identifying and cloning a gene by introducing cDNAs (prepared by copying a collection of messenger RNAs) into cells and selecting those that express a particular feature (e.g., colony-stimulating factor activity).

Fab fragments: The pair of identical fragments of an antibody obtained by cleavage of the molecule with the enzyme papain. Each fragment consists of an intact light chain and a like-size portion of the heavy chain. Each of the Fab fragments contains a specific binding site for the antigen. See Fc fragment; Figure 1.

Fc fragment: The third fragment (*See* Fab fragments) of an antibody obtained by cleavage of the molecule with the enzyme papain. The fragment consists of the remaining portions of the two heavy chains linked by chemical bonds, and it contains the component that binds cell-surface receptors. *See* Figure 1.

FDA: The Food and Drug Administration, the federal agency that sets standards for, and approves claims about, the efficacy and safety of drugs and the quality of foods.

fibroblast: A connective-tissue cell that synthesizes fibrous proteins, such as collagen. Fibroblasts are widely used in experimental studies of cell growth, gene expression, and cell proliferation.

fluorescence-activated cell sorting: FACS, a technique for the rapid, automated separation and sorting of cells in a population of cells that are identified by their having been bound by distinctive fluorescent chemical tags.

folic acid: A B vitamin, widely distributed in vegetable and animal foods, that is active as a coenzyme in many metabolic pathways in all cells.

Food and Drug Administration: *See* FDA.

formaldehyde: The very simple organic molecule H_2CO, commonly used to disinfect and to preserve biological specimens. It is also an active intermediate in cellular metabolism.

formate: The salt form of formic acid, the simplest organic acid (HCOOH). It is active in cellular metabolism.

gamma interferon: IFN-γ, a cytokine released by T lymphocytes of the TH1 type that is involved in cellular (inflammatory) immune responses.

G-CSF: Granulocyte colony-stimulating factor, the cytokine that stimulates the development of blood granulocytes. *See* colony-stimulating factor.

gene: A sequence of DNA that encodes a protein or an RNA; a unit of heredity at a specific place on a chromosome.

genome: The complete, single set of the genetic material of a cell or an organism; the single molecule of DNA or of RNA in some viruses.

glaucoma: A disease in which increasing pressure within the eyeball can damage the optic disk and lead to a gradual loss of vision.

glycopeptides: Compounds that contain a number of linked amino acids, to some of which one or more sugar residues are attached.

GM-CSF: Granulocyte/macrophage colony-stimulating factor, the cytokine that stimulates the development of granulocytes in the circulating blood and the scavenging macrophages throughout the body. *See* colony-stimulating factor.

G-proteins: A large class of proteins that bind and hydrolyze (break down) GTP (guanosine triphosphate). They function in signal transduction (*which see*) between the receptor in the membrane that receives a signal (from a hormone or growth factor) and the effector molecule that transmits the signal to still other molecules in the cell to initiate such processes as secretion, cell changes, and cell division.

granulocyte colony-stimulating factor: *See* G-CSF.

granulocyte/macrophage colony-stimulating factor: *See* GM-CSF.

GTP: Guanosine triphosphate, which resembles ATP (*which see*), with the base guanine replacing adenine. It serves an important role in signal transduction.

GTP-binding proteins: *See* G-proteins.

guanosine triphosphate: *See* GTP.

hairy-cell leukemia: A rare form of leukemia that is treatable with alpha interferon.

helper T cells: Lymphocytes of two varieties, designated TH1 and TH2. The TH1 variety participate in immune responses at the cellular (inflammatory) level; the TH2 variety stimulate B cells to produce antibodies.

hematopoiesis: The formation of red blood cells (erythrocytes) or, more generally, of all blood cells, including lymphocytes, granulocytes, eosinophils, macrophages, and platelets.

hemoglobin: The iron-containing protein in red blood cells that is responsible for the transport of oxygen.

histocompatibility complex: *See* MHC.

HIV: Human immunodeficiency virus, the virus responsible for AIDS.

hormones: Certain small or large molecules that carry regulatory signals from one cell or tissue to another.

hybridoma: A hybrid cell produced by the fusion of two cells that is useful for the production of monoclonal antibodies.

hydrocortisone: *See* cortisol.

IFN-γ: *See* gamma interferon.

IG-1, IG-2: Insulin-like growth factors.

Ig: Immunoglobulins (antibodies).

IgA: An Ig type that constitutes about 15 percent of the Ig in serum. It is the predominant Ig type found in bodily secretions.

IgE: An Ig type that constitutes less than 1 percent of the Ig in serum. It is associated with allergic reactions.

IgG: An Ig type that constitutes about 80 percent of the Ig in serum. It is the predominant Ig type produced in the secondary immune response.

IgM: An Ig type that constitutes about 5 percent of the Ig in serum. It is the predominant Ig type produced in the primary immune response.

IL: Interleukins, a family of cytokines designated IL-1 through IL-13, named in the order of their discovery.

immune complex: A complex formed by the binding of an antibody with an antigen.

immunogenicity: The capacity of a substance to induce an immune response; antigenicity.

immunoglobulins: *See* Ig.

IND: An application by a pharmaceutical company for the investigation of a new drug.

inflammatory bowel disease: *See* chronic bowel disease.

insulin-like growth factors: Substances resembling insulin that stimulate the growth of certain cells.

interferons: IFN, a family of regulatory cytokines that includes alpha interferon (IFN-α), beta interferon (IFN-β), and gamma interferon (IFN-γ).

interleukins: *See* IL.

Intron A: A trade name for alpha interferon.

in vitro: Pertaining to behavior outside a living organism; includes the use of extracts of cells, subcellular fractions, tissue slices, and perfused organs.

in vivo: Pertaining to behavior within a living organism; includes the use of a whole plant or animal or an intact microbial cell.

investigational new drug: *See* IND.

labile: Unstable, readily undergoing change.

Legionnaires' disease: A lobar pneumonia caused by the bacterium *Legionella pneumophila*; named after an outbreak at a convention of the American Legion.

Leishmania: A genus of flagellated protozoans that are parasitic in vertebrates and transmitted by insects. Human infections by *Leishmania* species include oriental sore and kala-azar.

Leucomax: A trade name for GM-CSF.

leukocytes (or *leucocytes*): White blood cells.

Lou Gehrig's disease: *See* amyotrophic lateral sclerosis.

lupus: Any of several diseases (e.g., systemic lupus erythematosis) characterized by skin lesions.

lymphocytes: The white blood cells (leukocytes) that include the B cells and T cells, but not the others (such as granulocytes and eosinophils).

major histocompatibility complex: *See* MHC.

mast cell: A large connective-tissue cell that can be stained by basic dyes. Granules in mast cells contain heparin and histamine, which are released in allergic reactions.

mast-cell growth factors: IL-3 and IL-4, cytokines that stimulate the growth of mast cells. *See* IL.

messenger RNA: mRNA, the single-stranded RNA synthesized during transcription and complementary to one of the strands of duplex DNA used as a template. It transmits the genetic information from DNA for translation into proteins by ribosomes.

metabolism: The sum total of all the physical and chemical changes that take place in a living system, which may be a cell, a tissue, an organ, or an organism.

methionine: One of the twenty common amino acids that constitute proteins; it is one of only two that contain sulfur.

methyl group: The CH_3 group attached to an organic compound.

MHC: Major histocompatibility complex, the chromosomal region containing groups of genes that encode the proteins controlling the rejection of transplanted (i.e., foreign) cells, tissues, and organs.

Micromonospora: A genus of actinobacteria, some of whose species produce useful antibiotics.

MRI: Magnetic resonance imaging, a euphemism, in medical circles, for NMR (*which see*), apparently to avoid the word *nuclear*.

mRNA: *See* messenger RNA.

MSTP: The prestigious Medical Scientist Training Program for combined Ph.D. and M.D. training in a select number of medical schools.

mutagenesis: Production of mutations by agents (e.g., x rays, ultraviolet light, certain chemicals) that damage DNA.

myositis: An inflammation of voluntary muscles.

National Institutes of Health: *See* NIH.

National Science Foundation: *See* NSF.

nerve growth factor: NGF, a cytokine, the first neurotrophic factor discovered. It is a protein that promotes the growth of processes from nerve cells and increases the metabolism of various nerve cells.

niacin: Nicotinic acid; a B vitamin that prevents or cures pellagra; a constituent of coenzymes that function in various essential metabolic operatioins.

NIH: National Institutes of Health, located in Bethesda, Maryland, with a federal budget in excess of $11 billion to support research and training in the medical sciences. It is composed of many categorical institutes (e.g., National Cancer Institute; National Heart, Lung, and Blood Institute; National Institute on Aging; etc.). The bulk of its budget is expended at universities and private research institutes all over the world.

NMR: Nuclear magnetic resonance, a spectroscopic technique that identifies molecules to atomic resolution for basic chemical studies and, clinically, for the diagnosis of disease. *See also* MRI.

norethindrone: A synthetic progestational hormone used in oral contraceptives.

NSF: National Science Foundation, the federal agency that disburses nearly $3 billion extramurally to support education, advanced training, and research in the physical and biological sciences and engineering.

nucleoside: One of several natural compounds made up of a purine base (adenine or guanine) or a pyrimidine base (uracil, cytosine, or thymine) linked to a sugar (ribose or deoxyribose). With phosphate attached to the sugar, the nucleoside becomes a nucleotide, one of the building blocks of RNA or DNA.

nucleoside analog: A compound, generally synthetic, resembling one of the known nucleosides altered either in the base or in the sugar.

nucleotide: A building block of RNA or DNA, consisting of a nucleoside with an attached phosphate. A long chain of nucleotides is known as a polynucleotide (nucleic acid), or, more specifically, as RNA or DNA. A short chain of nucleotides (i.e., under ten units) is known as an oligonucleotide.

Ocusert: A trade name for an intraocular device for the controlled delivery of a drug, such as pilocarpine, to the eye. *See* Figure 2.

Okayama-Berg procedure: An improved method for preparing intact cDNA for expression cloning. *See* cDNA; expression cloning.

oligonucleotide: *See* nucleotide.

oncogene: One of more than fifty such genes, generally found in RNA viruses, with a potential to cause cancer. Cellular genes (proto-onco-genes) are converted to oncogenes as a result of mutation or re-arrangement in the viral genome.

oxytocin: A hormone, consisting of nine amino acids, that is secreted by the pituitary gland and that causes contraction of smooth muscle. It is commonly administered (as Pitocin) to women in labor to stimulate delivery.

paper chromotography: *See* chromotography.

PCR: The polymerase chain reaction, in which DNA polymerase re-peatedly copies a segment of DNA, resulting in enormous amplifi-cation of a tiny sample. The technique is widely used in research, medical diagnosis, and forensic medicine.

pilocarpine: An alkaloid drug used in the treatment of glaucoma.

plasmid: An extrachromosomal, circular DNA found in bacteria, often containing genes that confer resistance to antibiotics, toxic metals, and poisons. Plasmids are used in genetic engineering to introduce a foreign gene that, when expressed, makes the bacterium a factory for the product of that gene product (e.g., human insulin, growth hor-mone, cytokines).

polymerase chain reaction: *See* PCR.

polyoma virus: A small DNA virus that produces tumors in rodents.

progesterone: The major female sex hormone, a steroid secreted by the ovary that is required for the maintenance of pregnancy.

Progestasert: A trade name for an intrauterine device for the controlled delivery of progesterone to the uterus.

proteins: Chains of twenty distinctive amino acids in a particular se-quence of one hundred or more that perform virtually all cellular functions, including those of enzymes, receptors, cytokines, interfer-ons, and antibodies.

Proteus: A genus of aerobic, gram-negative bacteria, some species of which are associated with human urinary-tract infections.

Pseudomonas: A genus of short, rod-shaped bacteria that includes sap-rophytes (living on dead and decaying organic matter) and animal and plant pathogens.

psoriasis: A chronic skin disease characterized by circumscribed red patches covered with white scales.

quinone: A benzene molecule with two attached oxygen atoms (dioxy-benzene) or some derivative of such a molecule. Quinones occur nat-urally in compounds that serve in the transfer of electrons and as components of synthetic drugs.

***ras* oncogene:** One of a family of genes first found in the retrovirus caus-ing sarcoma (muscle cancer) in rats.

reagent: A substance that participates in a chemical reaction. Reagents are commonly used to detect or determine another substance.

receptor: A molecule on the surface (membrane) of a cell that serves as a target for the binding of a hormone, drug, virus, or other entity that triggers a physiological or pharmacological response.

RNA: Ribonucleic acid, the many forms of which include messenger RNA (*which see*), transfer RNA and ribosomal RNA (used in the translation of messenger RNA), and viral RNA, which functions as the genome of certain viruses.

septic shock syndrome: An acute and often fatal clinical condition caused by infection, generally by gram-negative bacteria, leading to shock and multiple organ failure. No effective treatment is currently available.

signal transduction: The process whereby a signal delivered by a hormone, drug, virus, or other entity to a receptor (*which see*) on a cell surface is transmitted via G-proteins (*which see*) or some other means to cellular elements that trigger physiologic or pharmacologic responses (e.g., cell growth, nerve action, vision.)

somatomedins: A group of small proteins, released by the human liver and/or kidneys, that mediate the action of growth hormone on skeletal tissue and produce insulin-like effects in various target tissues.

somatostatin: A hormone from the hypothalamus that inhibits the release of growth hormone from the pituitary gland.

spectrophotometric assays: Quantitative analyses using the spectrophotometer, an instrument allowing measurements of selective absorption of light from the ultraviolet to the infrared ranges of the electromagnetic spectrum.

staphylococci: Bacteria of the genus *Staphylococcus*—spherical grampositive bacteria in pairs, tetrads, or irregular clusters—including parasites (e.g., *S. aureus*) of the skin and mucous membranes.

Streptomyces: Soil bacteria (actinobacteria, or actinomycetes), some of which produce important antibiotics (e.g., streptomycin).

substrate: *See* enzyme.

suppressor gene: A gene encoding a product that overcomes the action of another gene, commonly a mutated gene, when both are present.

T cells: A class of lymphocytes that includes helper T cells (*which see*) and suppressor (killer) T-cells.

TH1, TH2: *See* helper T cells.

thymidine: A nucleoside, made up of thymine and deoxyribose, that is found in DNA but not in RNA. It is commonly used as a radioactive tracer to measure the synthesis of DNA in cells and tissues.

tissue plasminogen activator: TPA, a protease (protein-cleaving enzyme) that converts plasminogen to plasmin and thereby causes blood clots to dissolve.

T lymphocytes: T cells (*which see*).

TNF-β: Tumor necrosis factor β, a cytokine.

transcription: The process whereby DNA is converted by enzymes to RNA (e.g., messenger RNA) so that the genetic information in DNA can be transmitted for the synthesis of proteins.

tryptophan: One of the twenty amino acids commonly found in proteins.

toxic shock syndrome: An acute and sometimes fatal disease caused by toxins released by bacteria (e.g., *Staphylococcus aureus*). It occurs almost exclusively in menstruating females using tampons.

Index

References in *italic* refer to illustrations.

A

Abbott, 123, 171, 205, 207
Abrams, John, 131
Acetaminophen, 252–53
Acetic acid, 179
Actinobacteria, 102
Activase, 201
Acyclovir, 10
Adams, Ray, 218
Adenosine triphosphate (ATP), 23
Adenovirus, 39
Adrenal gland, 62, 64, 69, 71, 164
Affymax, 93–94, 160, 174
Aging, 11, 15, 222
Aguayo, Albert, 222, 223
AIDS (acquired immune deficiency syndrome), 6, 8, 10, 11, 25, 73, 146, 215, 257
Alabama, University of, 148, 193
Alafi, Moshe, 111, 204
Alberta, University of, 42, 43
Alcohol, 217
Aldosterone, 164
Alemán Valdés, Miguel, *70*
Alien Property Custodian, U.S., 100
Alkermes, Inc., 175
Allen, Charles, 69, 74
Allergens, *139*
Allopurinol, 10
Alpha interferon, 112, 114, 117, 126, 129, 166, 201, 214, 233, 253
ALS (amyotrophic lateral sclerosis), 216, 226, 229
Alt, Fred, 223, 227
ALZA, 26, 28–31, 47, 53–56, 72, 73, 75–96, *83*, 99, 116, 124, 125, 193, 248
Alzheimer's disease, 216, 257

American Cyanamid Corporation, 110
American Heart Association, 185
American Society of Biological Chemists, 185
Ames, Bruce, 38
Amgen, 54, 111, 132, 141, 195, 202–10, 226, 230, 251, 256
Amikacin, 168
Aminoglycosides, 102, 111, 128, 168
Amorin, Carlos, 60
Ampex, 202
Amyotrophic lateral sclerosis. *See* ALS
Andreopoulos, Spyros, *173*
Anemia, 206, 208
Anfinsen, Christian B., 50
Angeletti, Ruth, 222
Angier, Natalie, 156
Antibiotics, 7, 102, 111, 113, 128, 129
Antibodies, 5, 10, 14, 15, 27, 28, 30, 47, *51*, 116, 131–34, *135–39*, 145–47, 150, 168, 227, 251; anti-idiotypic, 44; antidigoxin, 52, 53, 55; designer, 53; Fab fragments, 50, 51, 52, 55; Fc fragments, 50, 51; monoclonal, 10, 28, 30, 43, 50, 52–53, 54, 55, 248; polyclonal, 52–53, 55
Anticoagulants, 53
Antigens, 50, *51*, 52, 55, *135–38*, 154, 205
Antihistamines, 66, 101
Anti-inflammatory drugs, 64, 69, 73, 101, 150, 252
Antisemitism, 22, 96–97
Applied Biosystems, 54
Arai, Kenichi, 32–36, *33*, 45, 55, 133, 134, 140, 142, 153–55, 157, *183*, 227, 228–29, 249

Arai, Naoko, 32, 34, *35*, 36, 155
Arthritis, 10, 26, 63, 101, 150, 257
Arthur D. Little, Inc., 171
Asilomar conference, 180–81
Aspergum, 103, *106*
Aspirin, 101, 102–3, 252
Asthma, 90
Atlantic Richfield, 88
ATP (adenosine triphosphate), 23
Autoimmune diseases: and drug development, 10, 25, 27, 150, 251; and research, 47

B

B cells, 38, 43, 133, 134, *135*, 144, 146, 148–50, 188
Bacteria, 28, 38, 50, 111, 113, 133, 135, 137–39, 154, 179–80, 196
Baddour, Ray, 204
Baker, Charles A., 223
Baldwin, Robert (Buzz), 24, 221
Baltimore, David, 117, 227
Banchereau, Jacques, *144*
Barbasco, 65
Barbiturates, 53
Barcia, Alberto, 60
Barde, Yves, 222, 223, 225
Barth, Richard, 110, 120
Barton, Derek H. R., 164
Basic research, 7, 9–12
Bauer, Walter, 50
Bautz, Ekke K. F., 198
Bazan, Fernando, 157
BDNF (brain-derived neurotrophic factor), 225, 230
Beadle, George W., 23, 182
Bechtel Corporation, 57
Beclomethasone, 101
Becton-Dickinson, 54
Beecham, 256
Benjamini, Eliezer, 44
Bennett, Richard, 105
Benzodiazepine, 53
Berg, Paul, 2, 21, 24, 27–29, 47, 49, 55, 56, 127, 159, 162, 174–81, *177*, *183*, 184, 186, 194, 242
Berk, Arnold, 203
Berry, Harold, 117

Beta-Dart, 53
Beta interferon, 45, 112
Betamethasone, 101
Betaseron, 214–15
Bilirubin, 22
Binder, Gordon, 207
Biocine Company, 214
Biogen, 28, 107, 111–12, 116, 121, 123, 124, 127, 129, 204, 205
Biosynthesis, 11, 25
Bishop, J. Michael, 193
Black, Ira, 220
Bleomycin, 168
Blood-brain barrier, 216
Bloor, Walter Ray, 61
Blyth and Company, 202, 203
Boehringer-Mannheim, 251
Bonner, David, 182
Bowes, William K., 202–4
Boyer, Herbert, 28, 196, 198, 199, 207, 241–42
Bradshaw, Ralph, 222
Brain-derived neurotrophic factor. *See* BDNF
Braudel, Fernand, 4
Breast cancer, 11
Bristol-Myers, 105, 123, 168, 256
British Columbia, University of, 42, 72
Brown, Donald, 251
Brown, Francis C., 100, 102, 128, 164
Brown, Keith, 42
Brown, Michael S., 223, 227
Bugg, Charles, 148
Bullock, Frank, 115, 121, 124, 127, 160, 166, 169–72, *170*, 190
Burkitt's lymphoma, 146
Burroughs Wellcome, 73
Burton, Robert, 62

C

Calbiochem, 72
California, University of, at Berkeley, 38, 171, 202, 212, 240
California, University of, at Davis, 44
California, University of, at Los Angeles (UCLA), 203, 205

California, University of, at San Diego, 37, 188
California, University of, at San Francisco (UCSF), 48, 50, 193, 196–98, 200, 201, 203, 207, 209, 212, 213, 215, 241
California, University of, at Santa Barbara, 203
California Institute of Technology, 40, 203
Calvin, Melvin, 171
Cambridge University, 251
Cancer: and drug development, 10, 25, 26, 30, 107, 114–15, 117, 134, 168, 206, 214–15, 251; and immune response, 135; research, 9, 179–80
Cannabis, 217
Cantor, Harvey, 47, 48, 115
Carbon, John, 203
Carbon-14, 178
Carcinogenesis, 11, 236
Carlson, Chester, 8
Carnegie Institution, 202, 251
Carnegie-Mellon University, 252
Caruthers, Marvin, 196, 203, 207
Case Western Reserve University, 177–78, 182
Centocor, 54, 253
Cetus, 111, 129, 196, 201, 202, 204, 214, 234, 236–38, 240–41
Chao, Moses, 223
Chemokines, 157
Chess, Leonard, 115
Chicago, University of, 35, 203
Chiller, Jacques, 187–90, *189*
Chiron, 195, 201, 209–16, 240, 251, 256
Chlorpheniramine, 101
Chloroplasts, 42
Chlor-Trimeton, 101, 105
Cholesterol, 179
Ciba-Geigy, 28, 66, 88–90, 91, 99, 105, 110, 116, 119–21, 191–93, 214, 216, 226
Ciliary neurotrophic factor. *See* CNTF
Cisplatin, 168
City of Hope National Medical Center, 196, 238

Cline, Martin, 203
Clinical trials, 76, 82, 114, 129, 134, 151, 163, 168
CNTF (human ciliary neurotrophic factor), 226, 229, 233
Coccidial infections, 170
Coenzymes, 10, 23, 24, 40, 178, 179
Coffman, Robert, 32, 36–38, *37*, 134, *144*, 157
Cohen, Stanley, 28, 222, 241–42
Cohn, Melvin, 24, 37
Cold Spring Harbor Laboratory, 155–56
Cologne, University of, 193
Colony-stimulating factors (CSF), 132
Colorado, University of, 44, 188, 196, 203
Colostrum, 27
Columbia University, 35, 38, 115, 222, 228, 245
Commonwealth Scientific and Industrial Organisation (CSIRO), 42
Congress, U.S., 12, 16, 91, 120, 243
Contraceptives, 71, 73–74, 76, 79, 80, 82, 83–84
Conzen, W. H., 102–5, *104*, 107, 110–11, 116, 117, 164, 168
Cooper, Max, 193
Coppertone, 103
Cori, Carl, 23, 24, 152, 179
Cori, Gerti, 23, 24, 152
Coricidin, 101
Cornell University, 218, 220, 223, 224
Corticosteroids, 62, 75, 101, 164
Cortisol, 62
Cortisone, 62–64, 66, 69, 101
Cosmetics, 103
Cozzarelli, Nicholas, 35
Crohn's disease, 150
CSF (colony-stimulating factors), 132
CSIRO (Commonwealth Scientific and Industrial Research Organisation), 42
Cubist Pharmaceuticals, 175
Cyclamate, 91, 92
Cytokines, 5, 14, 15, 17, 33, 93, 132–33, *147*, *149*, 188, 208, 233; and

DNAX research, 132–34, 141, 142, 144–51, 154, 155, 156, 157, 163, 166, 172, 186, *187*
Cytorad, 53

D

Dalkon Shield, 84
Dana-Farber Cancer Institute, 48, 115
D'Andrade, Hugh, 3, 110, 114, 116, 118–24, *119*, 127, 159–60, *161*, 172, 190
Davidson, Norman, 203
DeKalb Company, 92
de Kruif, Paul, 5, 166
Delaney amendment, 91
Designer antibodies, 53
Diabetes, 9, 10, 215
Diazoxide, 101
Dig-Annul, 53, 126
Di-Gel antacids, 102
Digitalis, 52, 53
Digoxin, 52, 53, 55
Dioscorea, 65, *67*
Diosgenin, 65, 66, 68
Diuretics, 101
Dixon, Frank, 188
Djerassi, Carl, 26, 63–66, 69, *70*, 72, 74
DNA: amplification of, 214, 236–38; and animal viruses, 42, 179–80; and bacterial viruses, 179–80; cDNA, 133; and chloroplasts, 42; enzymology of, 10, 17, 25, 27, 29, 32, 180, 196, 199, 233, 236–37, 241; mutations of, 15; polymerases, 10, 25, 27, 73, 214, 233, 236, 240–41; repair of, 10, 27; replication of, 10, 11, 25, 27, 29, 32, 36, 45, 55, 149, 154, 180; SV40, 179–80; synthesis of, 11, 24, 25, 236–37; and viral chromosomes, 6. *See also* Recombinant DNA
DNAX: building, *122*; business plan of, 54–57; and cytokine research, 132–34, 141, 142, 144–51, 154, 155, 156, 157, 163, 166, 172, 186–87; founding of, 21, 26–31, 174–75, 181, 185–86; and IL-10 discovery,

133, 144–51, 153, 156; laboratory of, 31; and Schering-Plough, 2–3, 47, 57, 93, 107, 116–18, 121–27, 129–33, 140–41, 144, 148, 151, 153–55, 159–66, 169, 172–75, 186–87, 189–91, 193–94, 232; science plan of, 50–54, 93, 186–87; scientific advisory board of, 39, 47–49, 50, 193, 248; scientific staff of, 31–46, 142–43, 151–57, 193–94, 227, 229, 249
Dorfman, Ralph, 69
Dow, 256
Dreyer, William, 47
Drosophila, 42, 181
Drug-delivery systems, 74–87, 91, 93, 95
Dulbecco, Renato, 179
Du Pont Company, 15, 234, 236–38, 240, 243
Du Pont Merck Pharmaceutical, 201, 215
Dutton, Clarence, 9
Dyer, Rolla, 22

E

Eastman Kodak, 204
EBV. *See* Epstein-Barr virus
Ecdysone, 74
Edelman, Isidore, 35
Edison, Thomas, 233
Edmonton, University of, 156–57
Edsel Ford Research Institute, 166
Efstratiadis, Argiris, 227
Electron spin resonance, 47, 74
Eli Lilly, 71, 72, 123, 188, 198, 200, 213
Engenics, 29
Enzymes, 5–6, 14, 23, 40, 51, 60, 61, 152, 178, 179; and DNA processes, 10, 17, 25, 27, 29, 32, 55, 180
EPO. *See* Erythropoietin
Epogen, 206
Epstein-Barr virus (EBV), 146, 148
Erythropoietin (EPO), 30, 112, 133, 201, 203, 206, 207, 208
Escherichia coli, 25, 32, 37, 38, 45, 85, 180, 182, 184, 196

Estradiol, 100
Eukaryotes, 36, 45, 154

F
Fab fragments, 50, *51*, 52, 55
Fc-receptor, *138*, *139*
FDA (Food and Drug Administration),
 U.S., 52, 71, 79, 82, 84, 91, 93, 107,
 113, 120, 168, 214–15, 235
Federal Food, Drug, and Cosmetic
 Act, 120
Feen-A-Mint, 103, *106*
Fermentation, 13
Fiers, Walter, 121
Fierz, Marcus, 5
Fieser, Louis, 64, 66, 164, 170, 171
Flagellin, 145
Florida, Richard, 252
Florio, James J., *161*
Food additives, 91–92
Food and Drug Administration. *See*
 FDA
Fordyce, James W., 223
Fortanet, Don Emilio, 68
Free radicals, 74
Fried, John, 57, 218
Fruton, Joseph, 182
Fujizaki Pharmaceutical, 56
Furth, Mark E., 225, 228

G
Galactosemia, 210
GalaGen, 26, 27, 248
Gamma interferon, 146
Garamycin, 102, 111, 117, 128
G-CSF, 132, 141, 206, 207, 208
Gene expression, 38, 42, 50, 55, 180,
 185
Genentech, 28, 54, 73, 111, 123, 129,
 195–202, 204, 205, 207, 212, 215,
 227, 249, 256
General Electric, 205
Genetics Institute, 28, 54, 208
Genetics Society of America, 185
Gentamicin, 102, 128
Georgia, University of, 207
Gershon, Richard, 115
Gerstel, Martin, 76, 77, 82, 88, *89*, 93

Gibbons, James, 55
Gilbert, Walter, 112, 113, 121, 180, 198
Gilbert's disease, 22
Gilman, Alfred G., 217, 218, 220,
 222–23, 227
Glaucoma, 76, 82–83
Glaxo, 94, 101
Gleason, Frank J., 111, 121, 123-26,
 125
GM-CSF, 132, 134, 142, 163, 165, 233
Goeddel, David, 196–99, *197*, 202,
 227, 249
Goldby, Steve, 92
Golden Gate University, 54
Goldstein, Avram, 47, 74
Goldstein, Joseph, 223, 227
Goldwasser, Eugene, 203
Goodman, Howard, 198
Gordon Conference, 238
Gout, 10
Granulocytes, 132, 206
Greenberg, G. R., 178
Greene, Lloyd, 222, 223
Grobstein, Clifford, 211
Growth factors, 132, 134, 140, 144,
 154, 156, 204, 214, 217
Growth hormone, 28, 196, 198, 200,
 222
GTP (guanosine triphosphate), 33, 36
Gunsalus, Irwin, 127
Gurdon, John, 251

H
Haber, Edgar, 47, 50, 52–54, 55
Hall, Stephen S., 198–99
Hansen, Garth, 210
Harris, Lew, 171
Hartley, Brian, 112
Harvard University, 28, 47, 48, 61, 64,
 66, 115, 169–70, 198, 210, 243
Healy, Bernadine, 11
Hechter, Oscar, 62
Heidelberg, University of, 198
Heidelberg, Zentrum für Molekular
 Biologie in (ZMBH), 200
Hematopoietic system, 26, 93, 132,
 149, 157, 186, 233
Hemoglobins, 221

Hench, Philip, 63
Hepatitis, 114, 205, 213–14, 215
Heppel, Leon A., 23, 152
Herpes, 10, 25
Hershberg, E. B., 101, 105
Herzog, Hershel, 105
Hewlett-Packard, 202
Higuchi, Takeru, *83*
Hiser, Harold, 123
Histamine, 133, 134, *139*
Histocompatibility, 43
HIV (human immunodeficiency virus), 6
Hoechst-Roussel, 243
Hoffmann-La Roche, 236
Hofschneider, Peter-Hans, 112
Hogness, David, 24
Hood, Leroy, 39, 40, 47
Horecker, Bernard L., 23, 152
Hormones, 5, 15, 17, 26, 28, 33, 39, 53, 55, 111, 186, 209, 213, 217, 218; growth, 28, 196, 198; steroidal, 60–62, 65–66, 68, 69, 74, 75, 83–84, 100
Howard, Maureen, 143–44, *143*
Huddelson, Ed, 204
Human ciliary neurotrophic factor. *See* CNTF
Humoral response, 133, 145–46
Hunkapiller, Michael, 47
Hybritech, 188
Hydrocortisone, 62, 69, 75, 101
Hypertension, 26, 88, 91, 101, 126

I

IAP (intracisternal A particles), 146
ICOS, 195, 209
IgA, -E, -G, -M. *See* Immunoglobulins
IL-2, 146–47, 214, 215
IL-3, 134
IL-4, 134, 142, 146–47, 148, 163, 164, 172
IL-5, 134, 146–47
IL-6, 134
IL-10, 133, 144–51, *147*, 153, 156, 163, 172, 186
IL-13, 146–47

Illinois, University of, 54, 127, 203, 210
Immune system, 8, 10, 26, 28, 47, 93, 115, 124, 132–34, *135–39, 145,* 145–51, 186, 233
Immunex, 54, 132, 141
Immunoglobulins, 38, 40, 43, 47, 48, 134, 138–39, 146; IgA, 38, *139*; IgE, 38, 48, 134, *139*, 146, 148; IgG, 38, *138*; IgM, 38
Infectious diseases, 6, 27
Inflammation, 133, 135–38, 145–47, 149, 150, 186, 209
Insulin, 14, 53, 91, 196, 198, 200, 214, 222, 233
Interferons, 28, 30, 45, 112–15, 117, 124, 126, 129, 146, 166, 201, 213, 214–15, 233, 253
Interleukins, 132–34, 144–51, 213, 214–15
Intracisternal A particles (IAP), 146
Intrauterine devices, 84
Intron A, 114, 121, 166
Invention, process of, 8–9
Ishizaka, Kimishige, 47, 48, 146
Ishizaka, Teruko, 47, 48, 146
Isselbacher, Kurt, 47, 50
Itakura, Keiichi, 196

J

Jardetsky, Oleg, 170
Jaundice, 22
Jefferson, Thomas, 233
Jenkins, Lee, 111
Johns Hopkins University, 48, 146
Johnson, Franklin, 204
Johnson, Irving, 188, 200
Johnson, William, 66
Johnson & Johnson, 123, 188, 208, 214, 253
Josse, John, 73

K

Kafatos, Fotos, 227
Kaiser, Dale, 24, 179
Kalckar, Herman, 178–79
Kaplan, Henry, 54

Kapuscinski, Ryszard, 4
Kauzmann, Walter, 40
Kaziro, Yoshito, 32, 33, *33*, 34, 35, 36, 45, *49*, 112, 155
Kefauver, Estes, 120
Kendall, Edward C., 63
Kennedy, John F., 221
Kenyon College, 66
Keutmann, Henry, 62
Key Pharmaceuticals, 90
Khorana, H. Gobind, 72, 197, 236–38
Kleid, Dennis, 197
Kleiner, Eugene, 196
Kleppe, Kjell, 238, 240
Kleppe, Ruth, 238
Knight, Andrew, 56
Koch, Arthur, 37
Kogan, Richard, 105, 118, 127, 160, *161*, 172, 190–93, *192*
Köhler, Georges, 52
Korn, Laurence J., 251
Kornberg, Arthur, 19–29, *20*, 47, 49, 54–56, *83*, 127, 152, 159, 162, *173*, 174–75, 178–79, 184, 194, 223, 236–37
Kornberg, Ken, 31
Kornberg, Roger, 47–48
Kornberg, Sylvy, 24
Kornberg, Thomas, 47, 48
Kunkel, Louis, 223

L
Laboratorios Hormona, 65
Lagos, Licio, *71*
Lambda virus, 179
Lane, Alexander A., 105, 121, 123, 124, 127, 144, 160, 166–69, *167*, 172, 190
Lardy, Henry, 210
Lasker, Mary, 253
Lawrason, Douglas, 105, 168
Laxatives, 100, 103
Lederberg, Joshua, 72, 182, 184, 221
Lederle, 105, 110, 121
Lee, Frank, 32, 38–39, *39*, 42, 134, 157
Lees, Emma, 157

Legionnaire's disease, 11
Lehman, Robert, 24, 180
Leishmania, 146
Leprosy, 146
Lethers, David, 56
Leucomax (GM-CSF), 132, 163
Leukemia, 107, 114, 115
Leukine, 132
Leukocytes, 112, 132, 133
Levi-Montalcini, Rita, 222
Levinson, Arthur, 200
Levy, Ronald, 47
Liebig, Justus von, 13
Lilly. *See* Eli Lilly
Lindsay, Ronald M., 225, 228
Linn, Stuart, 240
Lipmann, Fritz A., 179
Liposome Company, 223
Little. *See* Arthur D. Little, Inc.
Litton Industries, 205
Lobban, Peter, 180
London, University College in, 221, 225
Lou Gehrig's disease. *See* ALS (amyotrophic lateral sclerosis)
Luciano, Robert P., 3, 57, 99, 102, 105, 107–18, *109*, 120–23, 129, 154–55, 159–60, *161*, 172, *173*, 190–93
Luedemann, George M., 101–2
Lupus, 10
Lymphocytes, 188
Lymphoma, 115, 146
Lynen, Feodor, 179

M
McConnell, Harden, 47, 74
McDevitt, Hugh, *173*
McGill University, 222
MacKinnon, Don, 192
McKnight, Steven L., 202
Macrophages, 132
Magnetic resonance imaging (MRI), 248
Maniatis, Tom, 227
Marion Laboratories, 84, 256
Marker, Russel E., 65–66
Martin, David, 201, 215

Martin-Marietta, 213
Massachusetts General Hospital, 47, 50, 52, 54, 126, 218
Massachusetts Institute of Technology (MIT), 174, 204, 227
Mast cells, 133, 134, 139
Matsumoto, Kunihiro, 153, 155–56, 157
Max Planck Institute, 200, 222
Maybelline cosmetics, 103
Mayo Clinic, 63
MBS (Minimum Binding Site) peptides, 53–54, 131
Medco Containment Services, 256
Memorex, 202
Memorial Sloan-Kettering Cancer Center, 225
Menotti, Amel, 168
Merck, 57, 66, 68, 83, 101, 123, 170, 190, 201, 213–14, 230, 256
Merrill Lynch, 100, 212, 223, 226
Messenger RNA (mRNA), 133, 185
Metacorticosteroids, 164
Metcalf, Donald, 132
Methionine, 178, 179
Meticortelone, 101
Meticorten, 101
Metrigen, 26, 27
MG-Extractor, 53
Michaels, Alan, 28
Michigan State University, 168
Microbial oil recovery, 203
Micromonospora, 102
Middleton, Fred A., 223
Millennium, 157
Milstein, Cesar, 52
Minimum Binding Site peptides. *See* MBS
Minnesota Mining and Manufacturing (3M), 204–5
MIT. *See* Massachusetts Institute of Technology
Miyajima, Atsushi, 36, 45, *46*, 140, 157
Miyajima, Ikuko, 45
Moffatt, John, 72–73
Monoclonal antibodies, 10, 28, 30, 43, 50, 52–53, 54, 55, 248
Mononucleosis, 146

Monsanto, 226, 243, 256
Montevideo University, 60–61
Moore, Kevin, 32, 39–40, *40*, 131, *143*, 144, 146, 148–49, 153, 157
Mosmann, Tim, 32, 42–43, *43*, 131, 134, *144*, 146, 153, 156–57
Motulsky, Arno, 203
MRI (magnetic resonance imaging), 248
Mullis, Kary B., 236–40, *239*
Multiple sclerosis, 214–15, 257
Murray, Kenneth, 112
Mutagenesis, 11, 38, 236
Myasthenia gravis, 53
Myositis, 150

N
Nagabhushan, T. L., 114, 144, 172
Nagata, Shigekazu, 112
Nagoya University, 156
Narula, Satwan, *183*
National Academy of Sciences, U. S., 185
National Cancer Institute, 225
National Institute of Arthritis and Metabolic Diseases, 164
National Institutes of Health. *See* NIH
National Science Foundation. *See* NSF
Nerve growth factor. *See* NGF
Nervous system, 26, 216–17, 220, 222–25, 228–29
Neupogen (G-CSF), 132, 206
Neurospora crassa, 23, 182
Neurotrophic factors, 217, 220, 221, 222, 225, 226, 229–30, 233
New York University, 23, 192, 202
NGF (nerve growth factor), 217, 220, 221, 222, 230
Niacin, 182
Nicotine, 217
Nifedipine, 91
NIH (National Institutes of Health), 11–13, 22, 23, 24, 50, 64, 127, 143, 151, 152, 170, 178–79, 212, 243, 245–47, 251, 258
Nishizawa, Yoshihiko, 222
Nitroglycerin, 90
Nixon, Richard, 96

NMR (nuclear magnetic resonance),170
Nobel, Alfred, 238
Nobel Institute, 210
Nobel Prize, 25, 48, 50, 52, 63, 179, 180, 182, 185, 218, 238
Nomura, Junichi, 222
Nomura, Masayasu, 35
Norethindrone, 71, 73
Novo Nordisk, 214
NSF (National Science Foundation), 127, 245, 246–47, 252
Nuclear magnetic resonance (NMR), 170
Nucleoside diphosphate kinase, 179
Nucleosides, 73
Nucleotides, 11, 24, 73, 180, 184
Nutrition, 5–6, 22, 178

O
Ochoa, Severo, 23, 25, 33
Ocusert system, 76, *78–79*, 82–83, 84, 85, 88
Office of Technology Assessment, U. S., 55
O'Garra, Anne, 157
Oklahoma State University, 177
Oncogenes, 12, 36, 156
Oncogene Science, Inc., 225
O'Neill, William P., 54–55, 124, 126
Opiates, 217
OROS systems, *81*, 85–86, *86*, 89
Ortho-Chiron, 214
Ortho Pharmaceuticals, 71
Osaka University, 156
Oshima, Yasuji, 155–56
Osteoporosis, 215
Oxytocin, 217

P
Pachter, Irwin, 168
Palade, George, 47, 48, *48*, 49, *49*
Paoletti, Claude, 56
Papa, Dominick, 105
Papain, *51*
Paris, University of, 115
Parke-Davis, 128, 168, 256
Parkinson's disease, 216

Pars, Henry, 171
Pasteur, Louis, 13, 115
Patel, Marilyn, 240
Patent and Trademark Office, U.S., 8, 233–35, 237, 240, 241, 242
Patents, 5, 13, 26, 45, 75, 88, 131, 214, 231–44
Paul, William, 143
Pauling, Linus, 221
PCR (polymerase chain reaction), 214; and patent litigation, 234, 236–38, 240–41
PDL (Protein Design Laboratories), 251
Pechet, Maurice M., 164
Pellagra, 5
Peltier, Bert, 168
Penhoet, Edward E., 209, 212, 213
Penicillin, 7, 102, 171
Pennsylvania State University, 65, 176
Peripatus, 42
Perkins, Tom, 196
Perlman, Preston, 105, 115
Perphenazine, 101
Peterson, D. H., 63
Pfizer, 91
Pharmaceutical Manufacturers' Association, 105, 111
Pheniramine, 101
Phosphates, 72
Phospholipids, 62
Photosynthesis, 171
Pickett, Cecil B., 190
Pilocarpine, 76, *78–79*, 82–83
Pincus, Gregory, 62
Pituitary gland, 217
Plasmids, 28, 128, 133, 134, 180, 196, 199, 241
Plough, Abe, 102–4, *104*, 107
Plough, Inc., 100, 102–5
Plum, Fred, 218, 223
Polio vaccine, 7, 10, 248
Polyclonal antibodies, 52–53, 55
Polymerase chain reaction. *See* PCR
Polymeric appendages, 91–92
Polymeric membranes, 76, 78, 81, 86–87, 91
Polypeptides, 47

Polysaccharides, 15
Prednisolone, 101
Prednisone, 101
Pricer, Bill, 24
Princeton University, 40, 204
Prince Ventures, 223
Procardia XL, 91
Progestasert system, 76, *79*, 80, 82, 83–84, 85, 88
Progesterone, 64, 65, 69, 76, 79, 83
Progynon, 100
Proleukin, 215
Prosser, Ladd, 127
Protein Design Laboratories (PDL), 251
Proteins, 6, 15, 27, 29, 33, 36, 50, 51, 111, 113, 132–33, *137*, 179, 218
Proteus infections, 102
Pseudomonas, 102
Psoriasis, 71, 73, 150
Ptashne, Mark, 197
Public Health Service, U. S., 22
Public opinion, 9, 15–16
Pyribenzamine, 66
Pyrophosphate, inorganic, 24

Q
Queen, Cary, 251
Quinones, 170

R
Raab, Kirk, 207
Rajewsky, Klaus, 193
Ras oncogene, 36
Rat brain-derived neurotrophic factor. *See* BDNF
Rathmann, George, 204–7, *206*, 209, 210, 251
Razdan, Raj, 171
Razzell, Bill, 72, 73
Recombinant DNA, 8, 10, 24, 25, 28, 29, 40, 54, 73, 93, 181, 203, 207, 227, 248; and Cetus, 237; and Chiron, 213; and DNAX, 55, 131, 133, 140; and Genentech, 196, 199, 200; and Schering-Plough, 107, 111, 113–14, 129, 133; and Stanford-UC patent, 241–42. *See also* DNA

Regeneron, 26, 196, 208, 217–30, *224*, 248, 249, 251, 255
Reichstein, Tadeus, 63
Renen-Dart, 126
Rennick, Donna, 32, 43–44, *44*, 131, 134, 144, 157
Repligen, 174–75
Research, styles of, 5–17
Research Institute of Medicine and Chemistry (RIMAC), 164
Retroviruses, 146
Rheumatoid arthritis, 10, 150, 257
Rhodes University, 42
Rhone-Poulenc, 256
Ribonuclease, 50
Ribosomes, 185
Richards, John H., 40
Rickets, 5
Rideal, Eric, 220
Riggs, Arthur, 196
RIMAC (Research Institute of Medicine and Chemistry), 164
RNA: enzymology of, 10, 25, 211; messenger RNA (mRNA), 133, 185; synthesis of, 25; transfer RNA (tRNA), 184–85; and viral chromosomes, 6
Robertson, Channing, 28–29
Roche Institute, 36, 129, 202, 214, 236–37, 240–41, 251, 253, 256
Rochester, University of, 21–22, 61–64
Rockefeller University, 65, 117, 182
Roman, Herschel, 212
Rorer Group, 256
Rosenberg, Barnett, 168
Rosenkranz, George, 63, 64, 65–66, 69, *71*, 72, 74
Rosenthal, Milton, 117
Roth, Frank, 105
Rothschild, Victor, 56
Rotman, Boris, 72
Royal Society, 185
Rutter, William J., 198, 203, 204, 209–14, *211*, 251
Ruzicka, Leopold, 65

S
Saccharin, 92

Sakami, Warwick, 178
Salk Institute, 36, 37, 179
Salser, Winston, 203, 205
Samuels, Leo, 62
Sanderling Ventures, 223
Sandoz, 243–44, 251
Sanger, Frederick, 73, 180
Sanofi Corporation, 56
Saraka, 100
Sarsaparilla, 66
Schall, Tom, 157
Schaller, Heinz, 112
Schering, Ernst, 100, *103*
Schering-Plough, 49, 91, 99–107, 110–18, 134, 164, 168–69, 190, 193, 204, 253; and DNAX, 2–3, 47, 57, 93, 107, 116–18, 121–27, 129–33, 140–41, 144, 148, 151, 153–55, 159–66, 169, 172–75, 186–87, 189–91, 193–94, 232; Unicet laboratory, 163
Schimke, Robert, 203, 227
Schimmel, Paul, 174–75
Schleifer, Leonard S., 217–20, *219*, 222–26, 228–29, 251
Schlessinger, Joseph, 202
Schwab, Martin, 223
Schwenk, Erwin, 101
Scientific cultures, 13–15
Scripps Research Institute, 188, 243–44
Scurvy, 5
Searle, 226, 256
Sebrell, William Henry, 23
Secrecy, 5, 13, 45, 231
Securities and Exchange Commission, U.S., 80
Seeburg, Peter, 197–98, 200
Selectomer, 53
Selye, Hans, *83*
Seneca, 17
Septic shock, 215
Shannon, James A., 12
Sharp, Phillip, 112
SHARPS, 171
Sheehan, John, 171
Shell, John, *83*
Shell Laboratories, 54
Shine, John, 197–98

Shizuoka University, 45
Shooter, Eric M., 217, 220–23, *221*, 227
Shultz, George, 57
Sieff, Marcus, 56
Signal transduction, 45, 149–50, 157
Sing, George L., 223
Skeggs, Leonard, 178
Sloan-Kettering Institute for Cancer Research, 63
Smith, Lloyd Holly, 50
Smith, William M., 45, 56
Smith-Kline, 123, 256
Snow, C. P., 13
Sokol, Herman, 168
Somatomedins, 214
Somatostatin, 196
Somlo, Emeric, 65, 66, 68, 69
Sperber, Nathan, 105
Spicehandler, Jonathan, 190
Squibb, 123, 256
SRI. *See* Stanford Research Institute
Stadler, Louis, 182
Standard Oil, 57
Stanford Industrial Park, 26, 72, 92, 162
Stanford Research Institute (SRI), 196–97
Stanford University, 24–26, 28–29, 36, 38, 41, 47, 54, 66, 72, 74, 115, 154, 162, 180, 182, 184, 203, 211, 221, 222, 241–43, 251, 252
Staphylococci, 102
Stargene, 238
State Department, U.S., 151
Stebbing, Nowell, 207
Sterling, Wallace, 182, 184
Steroids, 39, 61–66, 68–69, 73, 74, 101, 111, 164, 179
Stever, Guy, 127
Streptomyces, 102
Strober, Samuel, 47
Sugen, 195, 202
Sumitomo Chemical Company, 222, 223, 226
Suntory, 56
SV-40 DNA, 179–80
Swanson, Robert, 196–97, 200, 201

Swartz, Morton, 50
Sydney, University of, 42
Synalar, 64, 69, 71–72, 73
Synergen, 229
Syntex, 26, 57, 64–75, 76, 80, 101,
    160, 207, 220, 222, 232, 256
Syva Corporation, 74

T
Tabachnik, Irving, 105
Tabor, Herbert, 24, 152
Takeda Chemical Industries, 56
Targeted delivery, 53, 55
Tatum, Edward L., 23, 182, 184
T cells, 38, 49, 115, 124, 131–34, 135–
    36, 145–49, 154, 156, 157, 186, 188;
    helper T cell (TH1, TH2) para-
    digm, 145–48, 147, 186, 187
Tenin-Dart, 53
Terman, Frederick, 182, 184
Testosterone, 100
Texas, University of (San Antonio),
    36
TH1, TH2. See T cells: helper T cell
    paradigm
Theorell, Hugo, 210
Thoenen, Hans, 222, 223, 225
3M (Minnesota Mining and Manufac-
    turing), 204–5
Thrombo-Lys, 53, 126
Thymidine, 156
Thymus gland, 133
Timoptic, 83
Tishler, Max, 170
Tissue plasminogen activator. See TPA
Tissue-type response, 133, 146
Tjian, Robert, 202
TNF. See Tumor necrosis factor
Tokyo, University of, 32–33, 112, 153,
    154, 157, 168
Tomkins, Gordon, 212
Toray Company, 45
Toronto, University of, 208
Tosco, 203, 207
Toxic shock syndrome, 11
TPA (tissue plasminogen activator),
    201–2, 233
Tranquilizers, 101

Transdermal drug delivery, 86–88,
    87, 89, 90
Transfer RNA (tRNA), 184–85
Transplantation, 10, 150
Trichlormethazide, 101
Trimeton, 101
Tryptophan, 182, 184–85
Tuberculosis, 11
Tugentman, Abraham, 96
Tularik, 195, 202
Tumor necrosis factor (TNF), 146,
    147
Tylenol, 252–53

U
Ulcerative colitis, 150
Ullrich, Axel, 197–98, 200, 202
Umezawa, H., 168
Unicet (Schering-Plough) laboratory,
    163
Upjohn Company, 63–64, 68, 101,
    110, 123, 168
U.S. Venture Partners, 203
Utah, University of, 62, 210

V
Vaccines, 7, 8, 10, 11, 17, 28, 205, 209,
    212–14, 215, 216, 248
Vagelos, Roy, 57, 213, 230
Valenzuela, Pablo D. T., 209
Vapnek, Daniel, 207
Varian Associates, 74
Varon, Silvio, 222
Vigneaud, Vincent du, 178
Vincent, Jim, 205
Viral infections, 6, 25, 28, 30, 133,
    135, 136, 209, 214
Virginia, University of, 218
Vitamins, 5–6, 22, 23, 40, 100, 178

W
Waitz, J. Allan, 114, 121, 126–30, 128,
    140–41, 144, 160, 165, 172, 173,
    186–87
Wakunaga Pharmaceuticals, 56
Wallace, Bruce, 238, 240
Walter, Carroll, 84
Warner-Lambert, 192, 256

Washington, University of, 203, 211–12

Washington University (St. Louis), 23, 24, 152, 179, 184, 213, 221–22, 243

Waterman, Robert E., 100, 102

Watson, James, 155, 199

Wayne State University, 69, 166, 168

Weinshenker, Ned, 92

Weinstein, Marvin, 101, 102, 105

Weissman, Irving, 36, 38, 43, 47, *48*, 115, 203

Weissmann, Charles, 112, 129

Weizmann Institute, 56, 96

Wesley-Jessen, 105

Wessels, Norman, 211

Western Reserve University, 177–78, 182

Whitcome, Philip, 205

Whooping cough, 215

Wigler, Michael, 155–56

Williams, Richard, 203

Williamson, Alan, 43

Wisconsin, University of, 35, 66, 184, 210, 236

Wolfe, Ralph, 203

Wolstadter, Sam, 203–4

Wood, Harland G., 178–79, 182

Wood, W. B., 47

Worcester Foundation, 62, 64, 69

Woyciesjes, Americo, 102

Wyden, Ron, 243

X

X rays, 248

XOMA, 26, 27, 248, 253

Y

Yale University, 48, 115, 182

Yamaichi Securities, 56

Yamanouchi, 251

Yancopoulos, George, 227–29, *228*, 249

Yanofsky, Charles, 2, 21, 27–29, 38, 41, 42, 47, 49, 57, 127, 159, 162, *170*, 174–75, 179, 181–86, *183*, 194, 251

Yeast, 13, 45, 156, 213, 214

Yokohama University, 34, 155

Yokota, Takashi, 36, 45, *46*, 140, 153, 156

Z

Zaffaroni, Alejandro (Alex), 59–64, *63*, 65–94 *passim*, *67*, *70*, 71, *71*, *77*, *83*, 94–97, *95*, *125*, 209; and Affymax, 93–94, 160, 174; and ALZA, 26–31, 47, 54–56, 73, 75–96, 99, 116; and DNAX, 2–3, 26–31, 36, 42, 47, 54–57, 93, 116–17, 121–23, 125–26, 131–32, 159–60, 166, 174, 186, 191, 232; and Dynapol, 91–92; and Syntex, 57, 64–75, 76, 80, 160

Zentrum für Molekular Biologie, Heidelberg (ZMBH), 200

Zlotnik, Albert, 144

Zoecon, 74

Zurawski, Gerard, 32, *41*, 41–42, 134, 157

Zurich, University of, 112